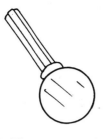

MORE

SCIENCE PROJECTS

YOU CAN DO

By George K. Stone

Illustrations by Mel Hunter

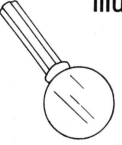

PRENTICE-HALL, INC.
Englewood Cliffs, N.J.

More Science Projects You Can Do, by George K. Stone

Library of Congress Catalog Card Number: 78-102399

Printed in the United States of America • *J*
0-13-600924-7; 0-13-600916-6 pbk.

PRENTICE-HALL INTERNATIONAL, INC., *London*
PRENTICE-HALL OF AUSTRALIA, PTY. LTD., *Sydney*
PRENTICE-HALL OF CANADA, LTD., *Toronto*
PRENTICE-HALL OF INDIA PRIVATE LTD., *New Delhi*
PRENTICE-HALL OF JAPAN, INC., *Tokyo*

Current printing (last digit):
10 9 8 7

• TO THE READER

This book is intended to help girls and boys decide upon an interesting project for the school science fair. As in the companion volume, *Science Projects You Can Do*, every project can be developed into a superior exhibit.

The projects do not require a textbook knowledge of science or special equipment. All can be done at home or in school, and the supplies needed can usually be found in a kitchen, dime store, or any number of other places.

Regardless of what interests you—animals, plants, astronomy, electricity, magnetism, optics, or almost anything else—you are likely to find a science project here that is ideal for you. Leaf through this book and you may come across a project you like, or consult Contents for a topical listing of projects.

This book differs from other science books you may have read because it is open-ended. This means that there is always something left for you to do or find out for yourself. But enough information is given so that you can proceed on your own. If you were to be told exactly how to do a science project, cookbook fashion, it would not be a science project at all and would not belong in a science fair.

Some of the ideas suggested in the following pages require original investigation. The answers are not known by anyone. In this sense the project will represent real research. These projects are in the area of basic science.

Another type of project that you will find in this book involves the collection and interpretation of scientific facts. The facts may be found by direct observation, through library research, and by experiments designed to reveal the facts. Such projects are in the area of natural history.

A third type of science project requires some mechanical skill, such as the construction of a working model. In some cases the customary materials are not at hand and others must be substituted. Sometimes a design must be created or modified, and some ingenuity or inventiveness is required. Projects of this type are in the area of technology.

Here are some suggestions that should be helpful in choosing, planning, developing, and exhibiting your project:

(1) Select a project in which you are especially interested or one that requires a skill you already have.

(2) Make the project your own; at least something about it should be entirely original.

(3) Be sure that you can obtain the necessary materials, preferably at little or no cost.

(4) Begin work well ahead of the time the project is to be exhibited.

(5) Keep complete notes of everything you do and present the data you obtain completely and legibly.

(6) Arrange the exhibit so that everyone can see all parts clearly and read the labels easily at a distance.

(7) The exhibit should be sturdy enough to stand moving and should fit into the space available.

(8) Be sure your exhibit is accident-proof and safe from souvenir collectors.

(9) If the project is to be a working model, it should operate without special attention or danger of breakage.

(10) Remember that a good exhibit should tell a story that others can understand.

• CONTENTS

NATURAL HISTORY 45

MECHANICS AND TECHNOLOGY 79

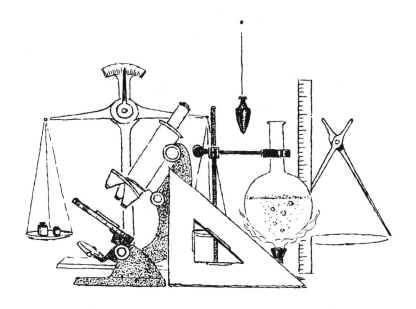

• BASIC SCIENCE

Basic science is the search for facts and principles that help to explain the world around us. New facts and principles are usually discovered by science projects of the kind suggested in this chapter.

• Say Hello to Yourself

Connect three or four 50-foot lengths of garden hose and coil them as shown in the diagram. Place your mouth against the end of the hose that connects with the faucet, and hold the other end against your ear. Now say *hello* quickly and loudly. The sound you hear arrives a fraction of a second later.

Many radio stations now carry "talk programs" in which there is a delay of several seconds between the sound that goes into the microphone and the sound that comes from the speaker. In case any language is used that the producer feels should not be broadcast, a key is pressed that interrupts the circuit. Find out how this delay is accomplished in a broadcasting station, and draw a diagram to show the circuit and hardware.

Could a working model be constructed by hooking up two tape recorders?

● Grow a Stalactite

A miniature stalactite will be formed if a piece of loosely woven cord or a narrow piece of lamp wicking is placed between two drinking glasses filled to the top with a concentrated solution of Epsom salts.

Dissolve as much Epsom salts as possible in warm water, and pour into the glasses. As the solution drips from the center of the cord or wicking, water evaporates and salt crystallizes out. If conditions are just right, the crystals will grow like a real stalactite. In addition, a stalagmite may form beneath it.

Experiment to find out how much the cotton strip must sag to make this happen. If the relative humidity is low, the stalactite will grow quite rapidly. If it is high, it will grow more slowly.

In caves the salt that is deposited as stalactites is calcium carbonate. This crystallizes out of solution much more slowly than magnesium sulfate (Epsom salts). Also, the relative humidity is always very low in caves where stalactites form.

In case you plan to use this idea in connection with a science project, you should be prepared with pictures of caves; a sample of a real stalactite if possible; and some understanding of relative humidity, capillarity, crystallization, and what is meant by a salt.

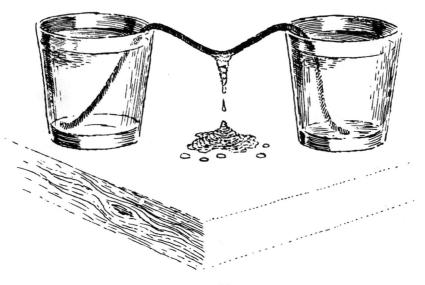

11

● Vortex Rings

Most amateur experimenters do not have technicians to make models for them. So they keep watching for items that can be salvaged to provide materials for their experiments.

Metal coffee cans are one of these useful salvage items, and they can be used for many kinds of experiments. Coffee cans are about 5½ inches tall and about 4 inches in diameter. They come with plastic covers that are also useful for many purposes.

One of the more interesting experiments that can be performed with a coffee can is the production of vortex rings. Use a can opener to remove the bottom of an empty can (the top has already been removed). Lay a penny in the center of one of the plastic covers and draw around it with a pencil. Carefully cut out the small circle and replace the cover on one end of the can. Place another plastic cover, with no hole in it, over the other end of the can.

Now we are ready to make vortex rings. Fill the can with smoke. There are several smoke-producing chemicals that make thick smoke when exposed to air. If you have difficulty in obtaining one of these, have someone blow tobacco smoke through a soda straw into the hole in the end of the can.

Lay the can sideways on an open book and lightly snap your finger against the end without a hole. A surprisingly small amount of energy is required to make the vortex smoke rings.

Now set a lighted candle several feet from the can and aim the vortex rings at it. What is the maximum distance at which the flame can be extinguished? Can the candle flame be extinguished equally well without the smoke?

Watch vortex smoke rings very carefully and try to observe the rotary motion within the rings as they travel.

With a supply of plastic covers, experiment with holes of different sizes. What diameter of hole seems to give the best results?

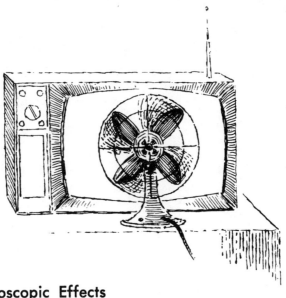

● Stroboscopic Effects

Set an electric fan in front of a television screen and turn on both the fan and the TV set. You should see the blades of the fan turning rather slowly either clockwise or counterclockwise. By turning the fan switch off and on it is possible to stop momentarily the apparent motion of the blades.

The reason for this phenomenon is that the light that comes from the television screen consists of many flashes per second. When the blades of the fan appear to stand still or turn slowly, the image of each blade has taken the place of the next one during the time that the light was out on the screen.

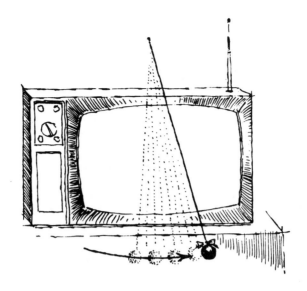

Something of the same effect is produced when you spread your fingers apart and wave your hand back and forth in front of the television screen.

Try tying a weight to a heavy cord and swinging it back and forth as a pendulum in front of the screen.

Find a spoked wheel, mount it on an axle, and spin the wheel. This will be more interesting if the wheel is spun with a rotator of some kind.

If you have a portable television set you may decide to set up an exhibit showing stroboscopic effects. In addition to the demonstration suggested here, you should be able to think of others. It may be necessary to do some library research in order to answer questions that visitors may ask.

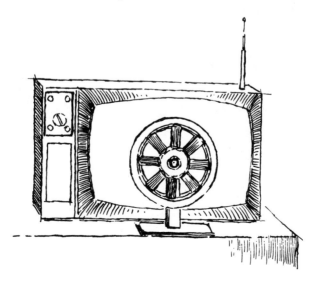

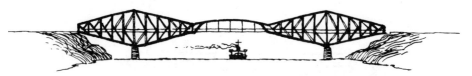

● Cantilevers

Long ago, before true arches were invented, the cantilever principle was used for supporting the walls above windows and doorways. Now it is used for many other purposes.

Place a stack of books, all the same size, on the edge of a table. Move the top book outward as far as it will go without tipping. Do the same with the next, and the next, and the next, until the top book is supported as shown below. Why is this possible? Look up the terms "cantilever" and "center of gravity" in reference books.

The principle of the cantilever would make an excellent topic for a science project with exhibit. In addition to the leaning stack of books, pile up two stacks so that they lean toward each other to form a cantilever arch. A model of a cantilever bridge could also be included. Be sure that you can answer the questions you might be asked about cantilevers.

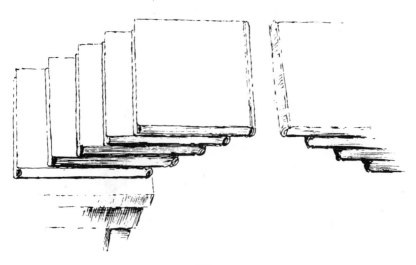

● Nails for Electrodes

Laboratory apparatus for the electrolysis of water and water solutions calls for the use of platinum electrodes, which are beyond the reach of most young experimenters.

However, experimentation has shown that stainless steel nails can be used in place of expensive platinum electrodes. The electrodes shown in the diagrams consist of No. 6 or No. 8 common stainless steel nails, pressed into holes which have been drilled into a solid rubber stopper. The stripped ends of copper lead wires are also fitted into the stopper beside the nails to seal off the solution. Press the stopper in to be sure the fit is tight.

For the electrolysis of water, ordinary washing soda (sodium carbonate) can be used instead of sulfuric acid, which is irritating to the hands. The apparatus can also be used to dissociate sodium chloride and other salt solutions.

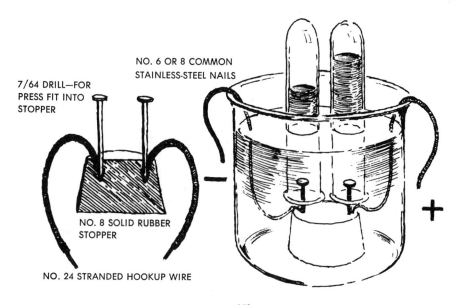

7/64 DRILL—FOR
PRESS FIT INTO
STOPPER

NO. 6 OR 8 COMMON
STAINLESS-STEEL NAILS

NO. 8 SOLID RUBBER
STOPPER

NO. 24 STRANDED HOOKUP WIRE

● Experiments with Mirrors

The numbers on the face of a clock run clockwise. But when we set a clock in front of a mirror and look in the mirror, the numbers seem to run counterclockwise. Why?

To help answer this question, tape two mirrors together at right angles and set the clock in front. The line where the mirrors are joined must run through the center of the clock when it is viewed in the mirrors. Now the numbers seem to run in the right direction.

When we look at the back of a clock, the numeral 3 is really on the left side, although we cannot see it. Light from the 3 is reflected into the mirror, which reflects it directly back to our eyes, as in the second diagram.

When two mirrors are set at an angle, light from the numeral 3 is reflected to the right-hand mirror, then to the left-hand mirror, and then to our eyes, as shown by the broken line on the diagram. The opposite is true for the numeral 9.

Take the clock away and look in the angled mirrors yourself. Adjust them so that the line joining the mirrors is even with the middle of your nose. Do you look natural? This is the way that others see you.

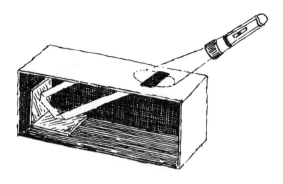

Two other examples of the reflection of light by two mirrors are shown on this page. At the top, two dime-store pocket mirrors are placed in a box with a slit cut in the side as shown. The reflected light beams can be seen against the other side of the box.

Set up the demonstrations on these pages and you should have a most interesting science project.

TWO POCKET MIRRORS

• Formation of Crystals

Add ordinary Epsom salts (magnesium sulfate), little by little, to a small amount of warm water until no more will dissolve. Stir in a few drops of liquid household glue. Then paint a small amount of the mixture on a piece of glass, such as a window pane. In a short time beautiful crystals will form. Look through a magnifying glass and watch them grow.

As the water evaporates, the molecules of magnesium sulfate, which are less than a hundred-millionth of an inch in diameter, arrange themselves into patterns somewhat as ice crystals do when they form on a window pane in freezing weather.

If you look closely at the crystals you will notice that many of them are not quite perfect. The imperfections are caused by small amounts of impurities—molecules that are not magnesium sulfate.

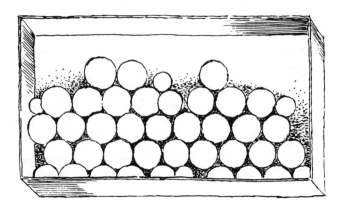

Actually, no chemical is ever absolutely pure. This will be easier to understand by using some marbles. Put the marbles in a rectangular pan and roll them all toward one side. If you shake them a little they will arrange themselves into a pattern with each marble touching six others. Now, if you push a larger marble in among them—an impurity—the pattern will be distorted. Something of the same thing happens to produce the beautiful colors in certain gemstones. The molecules are pushed out of place by the traces of impurities, and the gem therefore reflects light in a different way.

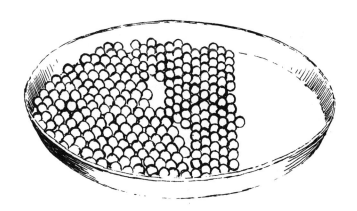

A second experiment will show another reason why few crystals are perfect. As a crystal grows, it is likely to meet another growing crystal. Where the two crystals meet they do not match up with each other.

Pour enough BB shot into the lid of a jar or can so that one side is covered. Then add more BB shot to the other side. The line where they meet is almost sure to be irregular. This is what happened to the Epsom salt crystals shown in the diagram on the opposite page. Crystals growing in one direction met crystals growing in another direction, and they did not fit together.

21

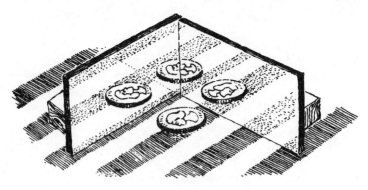

• Experiments with Reflected Light

If you set up two pocket mirrors at right angles and place a coin betwen them, you will see four coins. Let us see why.

First, it is necessary to understand what happens when the light from an object strikes a surface that reflects light.

Place some black liquid, such as water colored with India ink, in a pan. If you look directly into the pan from above at A you will see an image of yourself as in a mirror. If you want to see the square in the diagram, you will see it along the line B. If you want to see the ball, you will see it along the line C and nowhere else.

In the language of physics it is said that the angle of incidence equals the angle of reflection. This means that when light strikes a reflecting surface at a certain angle, it bounces off at the same angle.

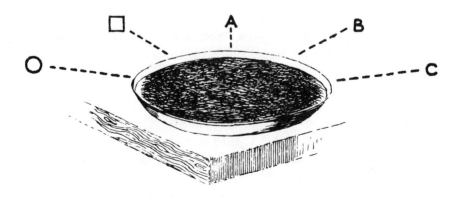

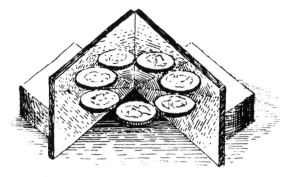

Now let us return to the pocket mirrors set at right angles. Move the mirrors so that you see a total of *seven coins*—six mirror images plus the real coin.

When you saw four coins it was because the light from the real coin was reflected by the right-hand mirror into your eyes. The image of this coin then acted as an object and was reflected to your eyes by the left-hand mirror, and the same thing happened in the opposite direction. The two halves of the remaining coin were also reflected in the same way. Now try to trace the reflections when you see seven coins.

Move the mirrors so that you see only *three* coins instead of four or seven. This will be easier if you put a piece of tape across the back where the mirrors join. Now try all other numbers up to ten.

This positioning of mirrors can be used to divide a circle into parts or laying off a fraction of a circle. Suppose, for example, that you are making a pie graph and want to find one-seventh of the circle. Set the mirrors as shown in the diagram above and draw lines along the bottom pie wedge. The result is one-seventh of a circle.

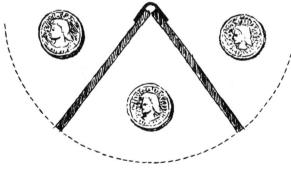

1/7th OF A CIRCLE

• Experiments in ESP

Experiment 1. Shown here are the symbols on ESP testing cards (Zener cards) widely used in tests for extrasensory perception (ESP).

Make a set of Zener cards from twenty-five squares of cardboard cut exactly the same size. The cards must be thick enough so that the symbols do not show through when held up to the light. Clearly draw on the cards these five symbols: cross, circle, star, square, and waves. Make five of each symbol—twenty-five cards in all.

Place five of the cards, one of each symbol, face up on a table. Then thoroughly shuffle the remaining twenty cards. Keep the twenty cards face down, and try to match them with the face-up cards. According to the laws of chance, you should be able to match correctly four of the twenty. If you can get six or more right and can do this several times, your ESP is working.

Now ask several others to try the same test, and keep records of their scores. Tabulate the results to determine (1) if some people get scores significantly above the statistical probability of four cards right and (2) if some people are more successful with certain symbols than they are with others.

This is only one of many tests that can be done with Zener cards. Consult reference books for further information and tests in the area of parapsychology.

Experiment 2. This test requires two people who will take turns testing each other. The two sit back to back, and each has pencil and paper.

First, the experimenter picks up and stares at the top card of a shuffled deck of twenty-five Zener cards. He tries to concentrate and "transmit" the image on the card to the subject. At a prearranged signal, such as the tap of a pencil, the subject indicates his decision about what card the experimenter was holding on the paper and the experimenter writes down what the symbol actually was. This is continued through all twenty-five cards. Then the lists are compared to find out how many "hits" there were. Any number above five is regarded as significant, but the test should be repeated at least five times and the number of "hits" averaged.

In another version of this test, the participants are in different rooms connected with some kind of electrical signaling device. If the results of these and similar tests are to be exhibited as a science project, include all the data in tabular form, as well as the cards and any other apparatus.

● More Experiments with Reflected Light

It is often impossible to use a movie projector or slide projector to good advantage in a small room. Also, the person operating the projector cannot point to features in the pictures when the screen is some distance away. Both of these difficulties are avoided by the use of a mirror, as shown in the diagram. This arrangement is often convenient as a supplement to a science project.

Set the mirror at an angle to the projector so that the reflected light beams are intercepted by it on their way to the screen. Chairs for the viewers should be arranged so that the light beams reach them at a similar angle. If a beaded screen is used, the pictures should be as clear as those projected in the usual way, and there should be little or no distortion.

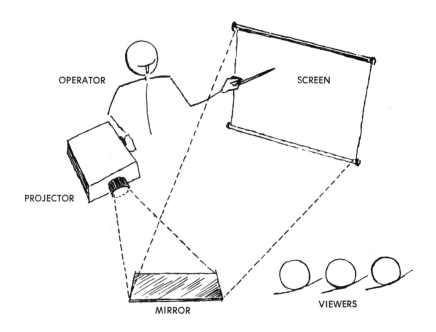

• Seeing Sound Vibrations

Obtain a metal can that has already been opened and remove the bottom with a rotary can opener. Stretch a piece of toy balloon over one end and fasten it with a rubber band.

Next you will need a small piece of mirror about the size of a dime. With extreme care this can be broken from a hand mirror by placing a piece of cloth over the mirror and striking it with a hammer. Put a drop of adhesive on the reverse side of the broken piece of mirror and press it against the stretched balloon, a little off center. Hold this device in the path of a beam of a projector so that a spot of light is reflected on a wall. If there is a sunny window in the room, sunlight can be used instead of light from a projector. Now sing different notes into the bottom of the can and watch the spot of light on the wall. A musical instrument can also be used for this experiment.

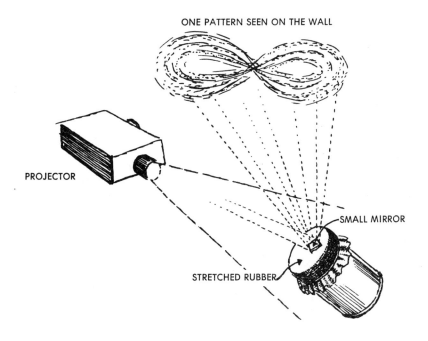

ONE PATTERN SEEN ON THE WALL

PROJECTOR

SMALL MIRROR

STRETCHED RUBBER

27

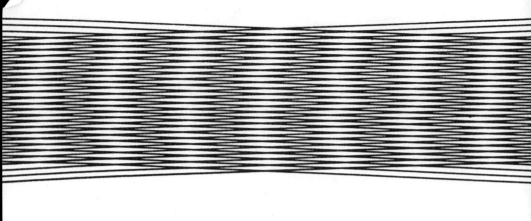

• Moire Patterns

Everywhere we look these days we see various forms of op art. One type, known as *moire patterns*, is made by having a series of closely spaced lines intersect at an angle—straight lines, curved lines, circular lines, or wavy lines can be used. The visual effect is based upon scientific principles that are applied in the production of fabrics, newspaper ads, movie and TV screens, and even to pictures appearing in art galleries.

The design at the top of this page is one of the simplest forms of moire patterns. It was made with a ball-point pen and a ruler. Anyone can do it by simply drawing parallel lines as close together as possible and intersecting these with another set at a slight angle. After that, with a little library research and with no special training in art, the designer can go on to the creation of more complicated and interesting designs.

Actually, moire designs are not new. They have long been used in making what is called "watered silk" and other fabrics. The "shimmer" in such fabrics is due to the small imperfections in the ribs that are made during the weaving process.

Today, special kits and transparencies are available for experimenters in this field, many of whom are now working in color. An exhibit of moire designs, both original and commercial, would make a most interesting science exhibit.

● Intermittent Funnel

Bend a piece of glass or plastic tubing into the shape of a question mark. Fit a one-hole rubber stopper over the straight end of the question mark and press the entire apparatus as far as possible into the neck of a funnel which has been set over a large bottle.

For demonstration and exhibit purposes it is best to use a glass or clear plastic funnel if you can find one.

Now slowly pour water (preferably colored) into the funnel. Watch the result closely and see if you and your viewers can explain what happens and why.

If you are planning to use the funnel as an exhibit, it would be interesting to have the water enter the funnel through a tube at a consistent rate.

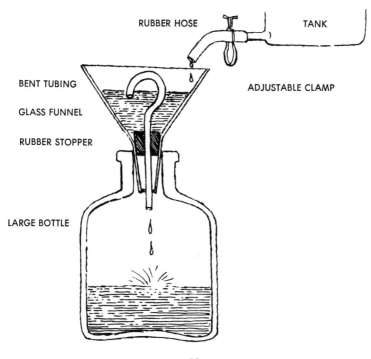

RUBBER HOSE   TANK

BENT TUBING   ADJUSTABLE CLAMP

GLASS FUNNEL

RUBBER STOPPER

LARGE BOTTLE

● The Bernoulli Effect

The most interesting exhibits at school science fairs seem to be those that show something in motion. (It also helps when the exhibit illustrates an important scientific principle.)

The Bernoulli effect is one of the important principles of science because it explains why the lift occurs when air flows over an airplane wing. The principle may be stated as follows: As the velocity of a gas or liquid is increased, the pressure perpendicular to the direction of flow is decreased.

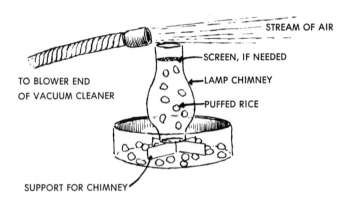

Arrange a tall lamp chimney so that it is supported about one-half inch above the bottom of a vessel containing puffed rice. Then direct a stream of air from the blower end of a vacuum cleaner across the upper end of the chimney.

Because the chimney is perpendicular to the flow of air, the pressure inside will be decreased. Grains of rice will therefore be pushed upward by atmospheric pressure as air enters the bottom of the chimney.

Some means of support must be devised for both the vacuum tube and the chimney. If suitable adjustments are made, the rice grains will bounce about inside the chimney, simulating the random motions of the molecules of a gas.

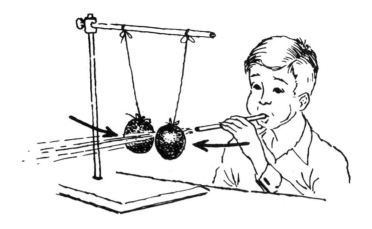

To accompany this exhibit, you might also have other demonstrations of Bernoulli's principle.

For instance, invite visitors to blow a stream of air through a soda straw between two suspended apples. Or suspend two sheets of paper so that they are about two inches apart and blow air between them.

Another demonstration can be done by bending down the ends of a 5 × 8 filing card so that the flat part of the card rests about an inch above a table. Invite visitors to blow under it through a soda straw. Or roll a 5 × 8 card so that it is curved. Set it upright and fan air across the convex surface.

How do these demonstrations illustrate the Bernoulli effect?

● Can Water Evaporation Be Reduced?

Every summer much water evaporates from lakes and reservoirs, and cities dependent upon these sources frequently are forced to impose water restrictions. Some engineers have suggested that the evaporation of water from lakes and reservoirs might be reduced by spreading some type of oil over the water surface. A simple experiment will illustrate the principle.

Get three plastic straight-sided refrigerator containers and put them in a place where they will not be disturbed. Fill each to the same level with water and measure the depth. Cover the

COOKING OIL ON WATER   OTHER LIQUID ON WATER   CONTROL (PLAIN WATER)

surface of the first jar with a thin layer of vegetable oil. Cover the surface of the second jar with a small amount of another liquid that will float. The third jar is the control—the surface should not be covered with anything.

Measure the depth of the liquid in each jar once each week for several weeks. Record the readings.

For the surface covering of the second jar, choose something without objectionable odor, and something that would be safe to drink. It has been found that sperm oil from whales reduces water evaporation in reservoirs. But this is scarce and expensive. Perhaps you can discover something better.

● More Experiments with Water

Make a hole in a vacuum bottle cork and fit a thermometer into it. Fill the bottle with very hot water and record the temperatures at regular intervals. Do the same with ice water in another vacuum bottle. Do the bottles gain and lose heat at the same rate in each case? Make a table showing the results.

Try the same experiment with some other liquid.

You are likely to find that liquids seem to lose heat faster than they gain it when placed in a vacuum bottle (thermos bottle). Why might this be true?

To test your hypothesis, you might proceed as follows: Place thermometers in two glasses of water, each of which is the same number of degrees above and below the room temperature. For example, if the room temperature is 70°F., the water in one glass might be 35°F. in one glass and 105°F. in the other. Record the temperatures in each glass as before, but this time they should be taken at shorter intervals, say every fifteen minutes. Make charts or graphs showing the data obtained from both experiments.

As an additional experiment with the same equipment, fill one glass with ice water and the other with ice water plus some ice cubes. Record temperatures at equal intervals as before. What conclusion can be drawn from this experiment?

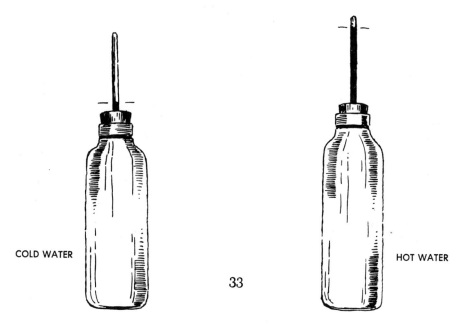

COLD WATER   HOT WATER

● Will the Glasses Overflow?

A good-size piece of ice is placed in a glass and the glass is filled to the brim with water. Will the water overflow from the glass when the ice melts? If you are not sure, try it and find out.

Another glass is filled level full with water. Salt is then shaken into it, a little at a time. The water is stirred very carefully with a thin wire to help the salt dissolve in the water. Does the salt seem to make the water level rise in the glass? How much salt can be added without overflowing the glass? To find out, measure the number of level teaspoonfuls that are put in the salt shaker and the number that remain in the shaker.

A third glass is filled with water, and rubbing alcohol is added little by little. How much can be added without making the glass overflow?

Scientists believe that there are spaces between water molecules into which molecules of salt and alcohol can fit without increasing the total volume. But there is a different reason why the glass containing the piece of ice does or does not overflow. Do you know what it is?

ICE

SALT

RUBBING ALCOHOL

• Making Water Wetter

Fill two drinking glasses with tap water. To the first, add a few drops of laundry or dishwashing detergent. The other glass of water is a control for this experiment.

Cut some pieces of cotton string, about half an inch long, and drop an equal number of them into each glass. In the first glass they will quickly sink to the bottom, but in the control glass they should remain on the surface of the water.

In the control glass the pieces of string float on the water surface because of the elastic water film on the water surface, called *surface tension.* In the experimental glass the detergent breaks this tension, which allows the fibers to get wet and the pieces of string sink to the bottom.

EXPERIMENTAL GLASS   CONTROL GLASS

Try this same experiment with other materials such as feathers, pieces of steel wool, talcum powder, sawdust, and ground black pepper. In each case, do they sink in water that contains a detergent and float on the surface of ordinary water?

To get a good look at the curved surfaces caused by surface tension, float a plastic berry basket, that has criss-cross ribs, on water. Such surfaces can also be seen clearly when a sewing needle or a safety razor blade is placed very carefully on water.

Detergents are often called wetting agents. They are helpful in laundering because they make water wetter and thereby loosen the dirt.

35

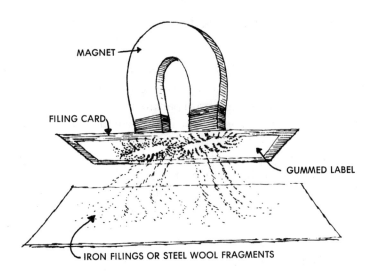

MAGNET

FILING CARD

GUMMED LABEL

IRON FILINGS OR STEEL WOOL FRAGMENTS

● Investigating Lines of Force

The usual way of demonstrating the lines of force around a permanent magnet is to lay a piece of glass on the magnet and sprinkle iron filings on the glass. Lo and behold, the iron filings arrange themselves into an attractive pattern.

But the magnetic field around a magnet is three-dimensional, as anyone knows who has ever played with magnets.

To get a clearer idea of the shape of this magnetic field you might proceed as follows: Sprinkle some iron filings on a sheet of paper. If you have no iron filings, pull apart a piece of steel wool so that little pieces of the steel fall on the paper.

Now paste the ungummed side of a gummed label, such as a parcel post label, on a filing card and paste the filing card (gummed label out) to the poles of a magnet. As soon as the paste holds firmly, moisten the gummed label. Hold the magnet with the card and label attached an inch or so above the paper with the filings or bits of steel wool on it. A magnet suspended in this way makes an excellent addition to an exhibit. Or when the paste is dry, the card can be removed from the magnet and turned filings side up. The gum on the label will hold most of the filings in place.

Cutting Magnetic Lines of Force

Suspend a piece of glass, such as a window pane, above a desk or table (A). Lay flat toothpicks on the pane of glass (B) and lay another piece of glass over them. The purpose of the toothpicks is to hold the panes a short distance apart.

Set a strong magnet (C) on the top pane of glass, and hold some carpet tacks (D) directly under the magnet and layers of glass. Even though the tacks are separated from the magnet, they are in its magnetic field and will be held against the glass.

Now run the blade of a steel knife (E) under the magnet *between* the two pieces of glass. The tacks drop off because the magnet force is deflected through the steel knife blade instead of through the tacks. Try other knives such as silver table knives and knives made from alloys. Try a piece of aluminum foil.

Think of a way to show the distortion of a magnetic field using iron filings, and test your idea.

Experiments with Polarized Light

Inexpensive sunglasses made of polarized plastic are sold in most dime stores and drugstores. Rotate two pairs of glasses as shown in the diagram and notice what happens. If you buy a pair and remove the two pieces of plastic, you have the material for many interesting experiments.

Using polarized light, look at objects you ordinarily would view with a hand lens. Devise ways to take pictures for an exhibit of polarized light.

Any microscope, even one of the inexpensive or "toy" microscopes, can be converted into a polarized-light microscope. Remove the eyepiece and unscrew the cap. Then cut a small circular piece from polarized plastic to fit inside. Also cut a larger piece and tape it on the stage of the microscope. Adjustments should be made by turning the eyepiece.

Some interesting things to observe are a double layer of cellophane, prepared microscope slides, the growth of crystals in a mixture of naphthalene and benzine, starch particles, various organic crystals, and slides of living microorganisms.

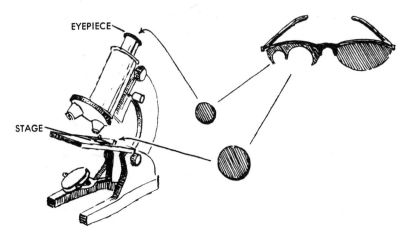

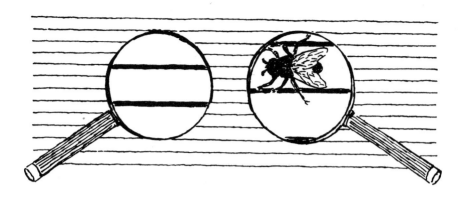

● Magnifying Power of a Lens

The number of times a lens magnifies or enlarges an image depends upon the curvature of the lens. A simple way of measuring the magnifying power of any lens is shown at the top of this page. Get the lines on a ruled card or piece of paper into focus. Notice the number of lines outside the lens and the number of lines under the lens and compare them. Count how many spaces are seen outside one of the enlarged spaces. In the diagram at top left there are three of these spaces, and the lens therefore magnifies three times, or 3×. How many times is the fly shown at the right magnified?

Set a filing card against the back side of a glass of water. How many times does it magnify when the lines are vertical? Horizontal? Explain.

These simple experiments and others that you can invent, or find in reference books under lenses, could be the basis of a very interesting exhibit.

● Effect of the Earth's Magnetic Field

For this experiment you will need a small magnetic compass and a can of food that has been sitting in an upright position for several days.

Hold the compass near the top of the can and note whether the N-pole or the S-pole is attracted. Now move the compass to the bottom of the can.

Turn the can bottom end up and test it again with the compass. What did you find? Leave the can in this inverted position for several days and test it again. For an exhibit, bring cans that have been sitting in each position for a week or more.

Make a list of other objects that seem to be magnetized naturally, such as radiators, stoves, refrigerators, steel files, door hinges, and iron floor lamps. Is the top of an automobile body magnetized in the same way as the lower part?

Since the earth has a magnetic field, an iron bar can be magnetized quickly by holding it in position and tapping it with a hammer. The hammering shakes up the iron particles so that they can line up more easily in the direction of the earth's field.

You can find out the best position to hold the iron bar by following the directions on the next page.

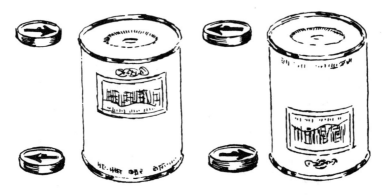

● Dipping Needle

The direction of the magnetic lines of force for any place on earth can be determined with the aid of a dipping needle. You can make one as shown in the diagram below.

Magnetize a steel knitting needle or a straight piece of steel wire by stroking it twenty or more times with a strong magnet. Run the magnet you have just made through a cork and press an unmagnetized sewing needle into each side as shown.

Now rest the unmagnetized sewing needles on the rims of two drinking glasses and adjust the apparatus so that the magnet and the needle bearings balance as perfectly as possible.

Set the glasses so that they are in an east-west direction and the magnet is in a north-south direction. The magnet should dip in line with the lines of force of the earth's magnetic field.

You will find that the magnetic lines of force strike the earth at a rather steep angle. That is why the objects mentioned above become magnetized, with the tops attracting the N-pole of the compass and the bottoms attracting the S-pole.

An excellent science exhibit could be set up by combining the dipping needle experiment and the experiment with the cans of food described on the opposite page.

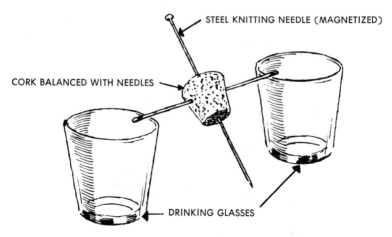

STEEL KNITTING NEEDLE (MAGNETIZED)

CORK BALANCED WITH NEEDLES

DRINKING GLASSES

41

• Floating Spheres

Place a little cooking oil in water and it floats on the surface. Drop it on rubbing alcohol and it sinks to the bottom. What will happen if the cooking oil is placed in a mixture of rubbing alcohol and water?

Partly fill a glass with rubbing alcohol and put a few drops of cooking oil on it. Add water, little by little, and try to get the drops to float under the surface. If too much water is added the drops rise (1). If more alcohol is then added, the drops sink to the bottom (2).

Mix some table salt with a little water to wash out the chemical that keeps it from taking up moisture. This is to keep the mixture clear when you repeat steps (1) and (2). Now the drops should float midway between the surface and the bottom of the glass (3). Why?

To make the drops stand out more clearly, add an oil-soluble dye such as violet methyl to the cooking oil (4).

Add just enough water to make one of the drops rise to the surface, and inject it with more oil from a sharp-pointed medicine dropper. Make the drops as large as possible, and then add a little alcohol to the mixture (5), (6).

The magnifying effect of the cylindrical drinking glass distorts the apparent shape of the drops so that they seem oval instead of spherical. To avoid this, select a straight-sided vessel and repeat the experiments. Try to get the colored oil drops as near the bottom of the vessel as possible (7). After a time, the oil drops rise to a higher level without the addition of either water or alcohol (8).

If you are presenting this demonstration as a science project, be sure you can answer such questions as: What was the purpose of putting salt in the bottom of the glasses? How does the density of the mixture in the bottom of the glasses differ from that at higher levels? Why do the oil drops become spherical in shape? Is salt soluble in rubbing alcohol?

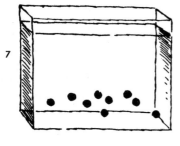 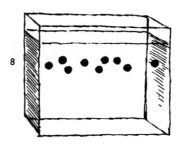

• Principle of Rockets

Dozens of model rockets are for sale in toy stores and described in books. None, however, are very suitable as working models at a school science fair.

The simple model shown here illustrates the principle behind all rockets. A sidearm test tube (A) is suspended from a bar by threads (B). A long piece of rubber tubing (C) is attached to the side opening of the test tube. Blow into the end of the rubber tube and the test tube will move in a forward direction as air escapes from the jet (D) in the mouth of the tube. Relate this to Newton's Second Law of Motion.

In case a sidearm test tube is not available, an empty dish-washing-detergent bottle makes a good substitute by sealing a tap into the side. Use your imagination to expand your rocket exhibit.

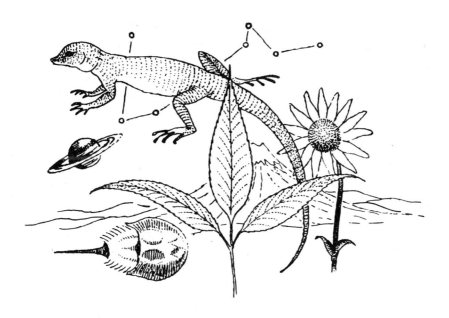

• NATURAL HISTORY

Natural history includes the study of animals, plants, rocks, soil, planets, stars, and everything else in nature. The methods of study are by direct observation and by experimentation, both controlled and uncontrolled.

Right-Pawed or Left-Pawed?

Most people are right-handed but some are left-handed. How about cats? If you look in books about cats, you are not likely to find the answer to this question. Possibly nobody knows the answer.

When a cat is hungry at feeding time, place some cat food in a glass container instead of in the regular dish. If the mouth of the container is too narrow for the cat's head, it will usually reach into the jar to try to get the food. Watch closely to see which paw the cat uses—the right or the left. Make a record of this. Repeat this for ten days or more and make a record of each trial.

In case your cat uses its right paw each time, all you can say is that "My cat seems to right-pawed." If it uses its left paw most of the time, it is a southpaw. If it uses each paw about the same number of times, or one paw and then the other, your cat is probably ambidextrous.

Do all cats behave as your cat does? To extend your research you will need more cats. Perhaps some friends can help you find out.

• Incubation of Reptile Eggs

Reptile eggs can be hatched as shown in the diagram. Place three or four inches of peat or moss in the bottom of an aquarium and dampen with water. Place the eggs on several layers of paper towels and set the thermostat at about 90°F. The aquarium should be kept in a dark room and the jar should be kept about three-fourths full of water. Watch the water carefully because it will evaporate quite rapidly.

Reptile eggs cost from 25¢ to $1 each and can be ordered by mail. Some reptile farms that supply eggs are: Southwestern Herpetological Research and Sales, Calimesa, California 92320; Snake Farm, Box 96, Laplace, Louisiana 70068; Zoological Supply Co., Laredo, Texas 78040; Thompson Wild Animal Farm and Zoo, RFD 2, Box 151, Clewiston, Florida 33440; Reptile Imports, Route 6, Box 199, Medina, Ohio.

Lizard and snake eggs are quite easy to hatch, but turtle eggs are more difficult. Be sure to read and ask questions about how to care for the young reptiles when they hatch.

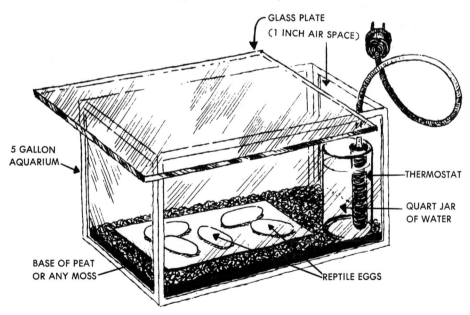

GLASS PLATE
(1 INCH AIR SPACE)

5 GALLON AQUARIUM

THERMOSTAT

QUART JAR OF WATER

BASE OF PEAT OR ANY MOSS

REPTILE EGGS

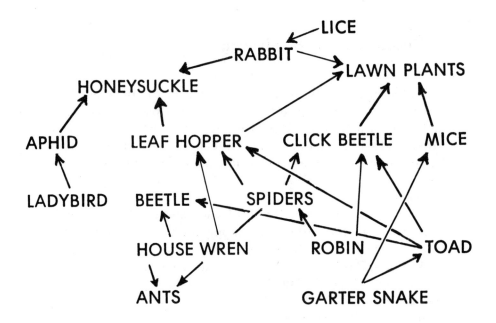

• Construction of a Food Web

The development of a food web that shows the complex relationships that exist in a biological community makes an excellent individual, team, or class project.

In such a diagram the arrows point toward the animal being eaten. Ideally, the diagram should be confined to a particular habitat, preferably to the local community, city, or country. Information should be based, at least partly, on direct observation, and checked by library research where any facts are in question.

A good way to start is by making a list of all the animals known to live in the local community, adding to the list as more are discovered. For the working diagram of the web, choose a large sheet of poster paper or cardboard so that there will be plenty of space to make additions as work on the project proceeds.

When the web is as complete as possible, revise it to improve the spacing. Names of plants and animals may be used as in the diagram here, but the final product will be more attractive if pictures can be found or drawn.

The first diagram shown here is obviously one made from the lawn habitat, and the other one shows a forest habitat. Obviously,

certain animals can cross from one habitat to another, producing interrelations between adjacent habitats. For instance, the garter snake from the lawn community is also an important member of the forested area.

It will be clear that no food web can include all the plants and animals in a given community. On the surface of the ground and below it may be found many kinds of fungi, including mushrooms, woody fungi, and molds. These provide food for various tiny animals usually impossible to identify. Some of them also provide food for larger animals.

This project, if it is done well, is not one that can be finished in a few days or a week. For best results and for greatest value it should extend over several months. As the seasons change, different animals will be seen at different times. Also, in any habitat there are likely to be migrating species that are seen for only a short time.

A project of this kind is a good example of original research, since it is unlikely that anyone has ever studied the living things of the selected habitat in this manner.

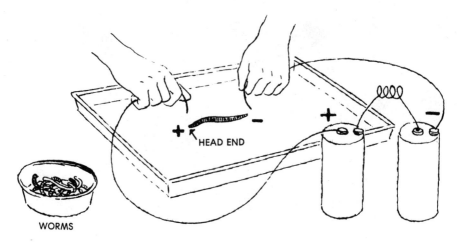

WORMS

• Polarity of Worms

Cover the bottom of a tray with several layers of moist paper. Place an earthworm on the paper as shown and hold the terminals from two dry cells as indicated. Does this affect the behavior of the earthworm and, if so, how?

Place the earthworm crosswise on the tray and place a terminal on each side. Again reverse the terminals. What happens, if anything? Repeat this experiment with at least ten worms and record the results.

Try the same experiment with other kinds of worms such as bloodworms and planarians.

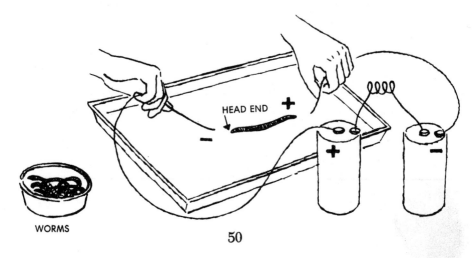

WORMS

• Responses to Light and Chemicals

For an hour or more, cover a tray containing several worms so that the worms are in the dark. Then shine a flashlight on them. Do they respond to light, even though they have no eyes?

Dip cotton swabs in different chemicals such as household ammonia and vinegar. Hold the swabs near, but not touching, a worm. Note any reactions. Repeat this with several worms.

Does it make any difference whether the swabs are held near the head end or the tail end of a worm? Which seems to be more sensitive, the head or the tail end?

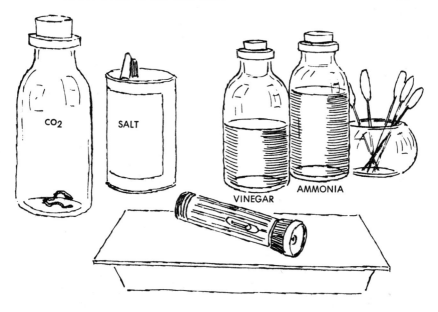

Place a container with two or more worms in the refrigerator for an hour or more, and repeat the experiments described on the opposite page. Are the results the same as those previously obtained?

Repeat the experiments described at the top of this page while the worms are still cold and make a record of your observations.

Insects can be "put to sleep" by placing them in carbon dioxide for ten or fifteen minutes. In a short time they usually revive. Is this also true of earthworms? Try it and see.

• Exploring a Strange Habitat

Many plants, such as those of the pineapple family, have leaves that catch rainwater. The water runs to the base of the leaves and remains for a time in little pockets. Push a piece of glass tubing into one of these pockets, put your finger on the upper end of the tube, and lift out a little of the liquid. Place a drop of this on a microscope slide and you should see many tiny living things moving about.

In dry weather there may be no liquid in the pockets. If there is none, pour some water on the plant and wait two or three days. Now draw out a sample, place a drop of it on a slide, and put a cover slip on it. Any inexpensive microscope can be used to examine the sample, even a "toy" microscope.

Scientists know little or nothing about the plant and animal communities that exist in these strange places. How do the tiny living things survive in dry weather? Are they of any use to the host plant? Do they obtain food from the host plant or in some other way? Where do they come from and how do they get there? These and other questions could keep a scientist busy for many years.

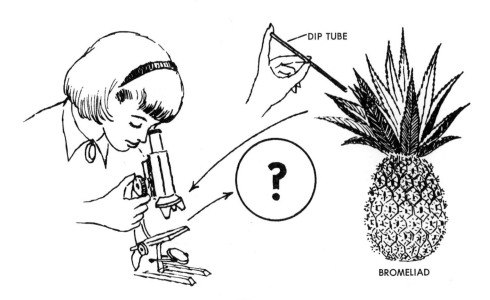

DIP TUBE

BROMELIAD

• Exhibit of Air Plants

Air plants can be found wherever you live. Some, such as lichens, grow on rocks. Others, such as the Bromeliad shown on the opposite page, usually grow on trees. Spanish moss is common through the southeastern states, and Usnea moss is found along the North Atlantic coast. Various other kinds are common along the Pacific coast, as well as inland.

Although most air plants grow on trees, they use the trees only for support. They take no food or water from them and usually do not harm them in any way. All true air plants manufacture their own food and obtain water from the air through spongy roots or modified leaves.

Like Bromeliads, other air plants provide homes for small animals, especially insects, spiders, and snails. If you pull apart some Spanish moss and shake it over a container you will have an interesting collection for an exhibit. The same is true for Usnea moss and similar air plants.

Air plants are well-suited for an exhibit since most people know very little about them. With the exception of lichens, most other air plants have tiny flowers and bear seeds. If possible, the exhibit should contain one or more orchids, since orchids are such attractive air plants. Where few materials are available locally, pictures can be used to supplement the exhibit.

ORCHID

SPANISH MOSS

● Model of the Zodiac

Everyone has heard about the signs of the Zodiac, but hardly anyone has a clear idea of what the Zodiac is or where it is. A good model of the Zodiac might help to remedy this situation.

The Zodiac is a belt of twelve constellations, or star-groups, that circle the sky close to the earth's orbit around the sun. Each month, a different constellation is directly opposite the earth on the other side of the sun.

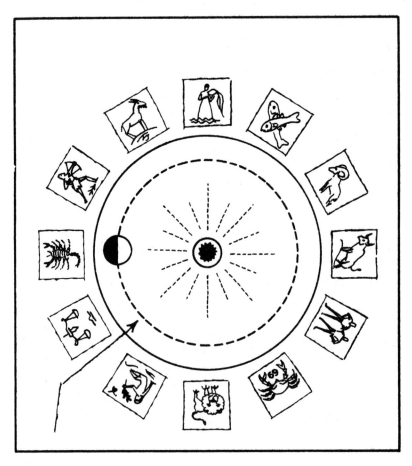

Draw freehand pictures of the twelve signs of the Zodiac on cards and lay them in proper order in a circle. If you are planning an exhibit, paste the cards on a very large sheet of cardboard.

Next, cut a smaller circle of cardboard and draw diagrams of the earth and sun on it. Push a tack through the center of the sun into a thin piece of wood so that the circle can be turned. Blacken the outer half of the earth to represent night, and set this circle in the center of the circle of cards.

According to the diagram on the opposite page, Taurus is opposite the sun. We say that "the sun is in Taurus." After a month, the earth will have traveled one-twelfth the way around its orbit. The sun will then be in Gemini, and so on around the other signs of the Zodiac.

This model also illustrates why we see different stars in summer than we do in winter. In the position shown, only the star groups on the left are visible from the dark side of the earth. All the other stars are shining, but we do not see them because the sun's light blots them out.

Turn the center disk in different directions and notice which star groups are visible at night and which are not.

If you should undertake a project of this kind, your exhibit might well include diagrams of other constellations besides the Zodiacal. In any event, some library research will be helpful and perhaps necessary.

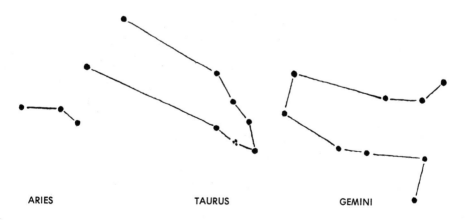

ARIES　　　　　　　　TAURUS　　　　　　GEMINI

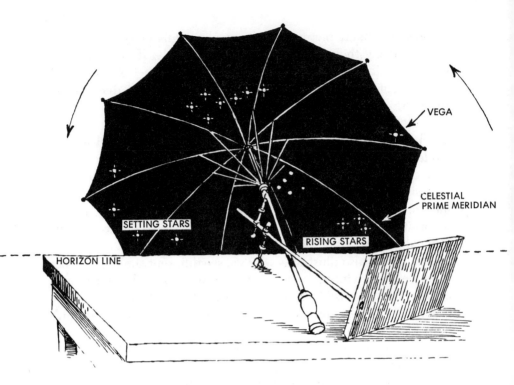

VEGA

CELESTIAL
PRIME MERIDIAN

SETTING STARS

RISING STARS

HORIZON LINE

● Umbrella Planetarium

A working model that shows the apparent motions of the stars in much the same way that large planetariums do can be made from a black umbrella.

Support the opened umbrella behind a desk or table at an angle to represent the inclination of the earth's axis. The umbrella is then presumed to point in the direction of the North Star.

In the diagram, a laboratory stand and two burette clamps provide this support, but wooden supports can be arranged to serve the same purpose. Adjust them so that the umbrella turns easily.

Now put some stars against the black sky. This can be done with white chalk or by pasting on bits of white paper. Start with W-shaped Cassiopeia, which is near the celestial prime meridian. Almost directly across from Cassiopeia is the Big Dipper, with the two pointers in line with the North Star.

Add more stars. Vega is usually the brightest one seen in the northern sky. Other stars can be put in at random because they are not important for what this planetarium is designed to show.

Slowly twist the shaft of the umbrella and watch Cassiopeia and the Big Dipper circle about the North Star. These stars do not rise and set because they are always seen above the horizon.

Watch Vega and the other stars set in the west below the horizon as the earth turns on its axis. Watch them rise in the east as the earth continues to rotate. Remember that the stars do not really move; they only seem to do so because the earth is turning.

Call attention to the ribs of the umbrella, which represent celestial meridians. Astronomers determine the celestial longitude of stars in the same way that the longitude of points on the earth are determined: so many degrees from the prime meridian. In the model shown, the meridians are thirty-six degrees apart.

Make a star finder to go with this model. Punch holes in a cardboard box to represent stars as they appear on a sky map. Cut a larger hole in one side for a flashlight, as shown. Now you have a sky map that you can read outdoors in the dark.

In case you plan to exhibit an umbrella planetarium in a science project booth, set it on a box so that viewers look upward.

FLASHLIGHT

• Working Model of a Water Table

Lowering of the water table has caused severe shortages of water in many places. The main causes are over-pumping, depletion of the underground water by drainage, and failure to provide sufficient replenishment of water from the surface. This can all be demonstrated with the model shown in the diagram.

Sand, gravel, and soil are put in a rectangular aquarium to represent the layers above the bedrock. (The bedrock is represented by the bottom of the aquarium.)

A piece of large-bore glass tubing is placed inside the aquarium against the glass to represent a deep well, and a smaller piece of tubing is placed at the opposite end to represent a shallower well. Between these, a depression is hollowed out to represent the cross section of a pond or lake. The land slopes upward toward the shallower well. If desired, tiny models of buildings and trees can be added to make the landscape look more realistic.

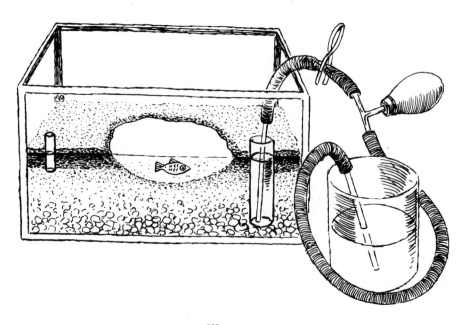

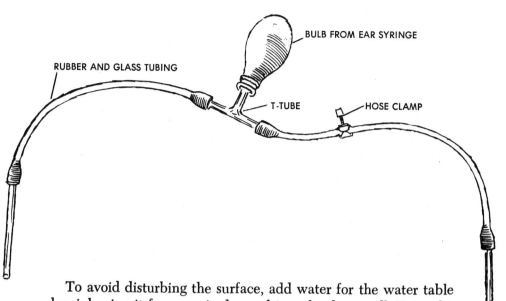

BULB FROM EAR SYRINGE

RUBBER AND GLASS TUBING

T-TUBE

HOSE CLAMP

To avoid disturbing the surface, add water for the water table by siphoning it from a raised vessel into the deep well. Raise the water level until it fills the pond and appears in the bottom of the shallow well.

The siphoning device shown here is convenient for adding water to the aquarium and operating the model. The purpose of the two pieces of glass tubing on the ends of the rubber tubes is to provide weight to hold them down in position. When adding water to the model, set the water supply on a box. With the clamp in position, squeeze and release the bulb to fill it. The siphon will flow in either direction, depending upon whether the water supply is to be raised or lowered.

Demonstrate the model by raising the water table to fill the pond and to allow water to rise in the shallow well. Then lower the water table to empty the pond and dry up the well. This is done by merely raising and lowering the vessel containing the water supply. To most viewers of your project, this will be both fascinating and instructive.

To add further interest to the model, whittle a small wooden fish, paint it, and weight it on the bottom so that it will just float near the water surface. When the pond is filled the fish will "swim," but when the pond is empty it will lie on its side on the bottom.

● Spider Web Collection

Spider webs are plentiful at certain times of the year and under certain weather conditions. Since no two are ever alike, they are interesting to collect.

When you find a good web specimen, get a sheet of black cardboard slightly larger than the circular part of the web, coat a narrow strip along the outside edge with glue, fast-drying adhesive, or shellac. Carefully press the cardboard against the web and snip off the supporting strands with scissors.

To preserve the web permanently, get a roll of plastic adhesive transparent film from a stationery shop. This is the material used to repair books and to protect drawings, documents, and photographs. Cut a piece the right size, peel off the backing sheet, and press the sticky film against the web.

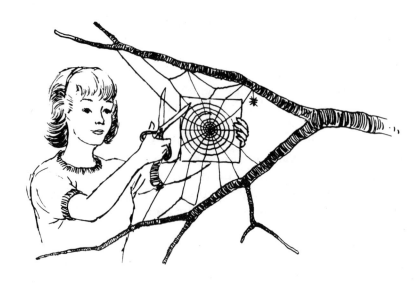

● Wild Flower Collection

Plastic adhesive transparent film is also ideal for preserving flowers if they are fairly small. Soon after the flowers are picked, press them between two pieces of blotting paper and cover these with a stack of heavy books.

The next day or later, lay the pressed flowers on white paper and cover them with the adhesive transparent film. Cut the film a little larger than the mounts so that the edges can be turned over neatly. Of course, the items in any collection should be labeled. Your collection should also be carefully labeled, including when and where the specimens were collected.

Specimens preserved by using this simple method may last many years in excellent condition.

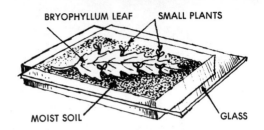

• Growing Plants without Seeds

Examples of various types of vegetative propagation make an excellent exhibit, especially at the time that most science fairs are held. Some of the examples, however, need to be started well in advance.

(1) Half bury a potato, stem down, in a pot of soil. Keep it moist in a good light, as you would a house plant, for three or four weeks.

(2) Lay a leaf of Bryophyllum, often called "sprouting leaf," on moist potting soil. Cover it with a pane of glass, except for a tiny crack, to keep the humidity high. Within four or five weeks new plants should start to grow along the borders of the leaf.

(3) Many bulbs will not grow roots when placed in water, but an onion will. Allow about three weeks.

(4) Soft-stemmed plants such as Coleus put out roots within about a week when slips are placed in water.

(5) Cuttings from woody plants are best started in sand.

Collect other plants that reproduce vegetatively, such as those that start by layering and from rhyzomes, corms, runners, and underground stems.

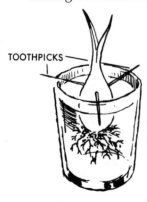

• Potato-Tomato Graft

During the regular growing season it is quite easy to graft a potato stem to a tomato stem so that potatoes and tomatoes are grown on the same plant.

Grow the potato plant and the tomato plant side by side, and when both plants are tall enough they can be grafted together as described below. However, this procedure does not work too well at the time that science fairs are held unless the experimenter has access to a greenhouse where the humidity is controlled.

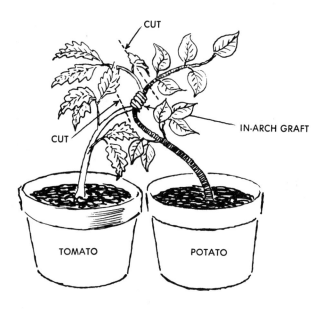

Grow the potato plant and the tomato plant in two large pots side by side, or together in a very large pot. When the plants are about a foot tall bring the two main stems together. Where they touch, shave the outer portion of each stem until several vascular bundles of each stem are exposed. Tie the cut surfaces together with soft cord and seal with grafting wax. Then cut off the stems as indicated in the diagram.

Make several grafts in this way and exhibit the best one, allowing about six weeks for results. When tomato fruits appear, carefully remove a little soil to show potatoes growing on the roots.

• Flower Model

Start with the flower stalk. Roll up a short piece of green modeling clay. Press one end against a flat surface and push a toothpick in the other end.

Fold a small piece of green paper on the diagonal three times. Holding on to the center, cut both ends to a point and place this over the toothpick to represent sepals.

Fold and cut brightly colored paper for a ring of petals and place this on the sepals.

Mold a pistil from clay and push this over the toothpick.

Make stamens from toothpicks and clay—yellow if possible. Push these around the base of the pistil.

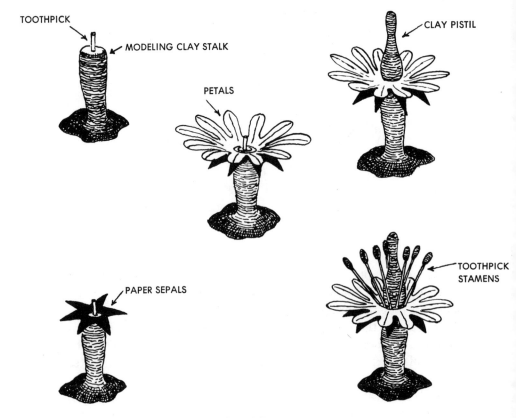

TOOTHPICK

MODELING CLAY STALK

CLAY PISTIL

PETALS

PAPER SEPALS

TOOTHPICK STAMENS

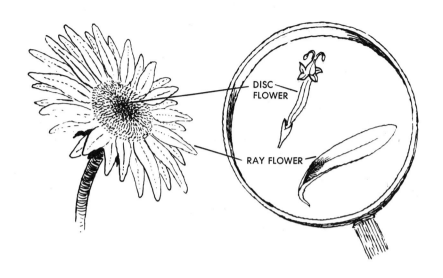

Hopefully, you have just made an excellent model of a simple flower, such as a lily or poppy. Of course, no two kinds of flowers are exactly alike. All differ in size, color, and shape of the parts.

Most of the flowers that we grow for decoration, as well as those that grow wild, belong to the composite family and are called *compound flowers*. These have a disk-shaped center surrounded by a ring of petals.

If you examine a compound flower with a magnifying glass, you will find that it is made up of a large number of simple flowers. Pull the center of a compound flower apart and you see a large number of tiny simple flowers, called disk-flowers. Examine one of these with a good magnifying glass and you will usually find the same parts as in the model of the simple flower.

Pull off some petals, or rays, and you will also find the same parts, except that there is only one petal.

For your exhibit, set up all five stages showing the construction of the simple-flower model. Include specimens of real flowers, both simple and compound, and provide lenses for visitors to examine them.

• Make a Frog Hibernate

A frog can be made to hibernate at any time of the year by gradually lowering the temperature of water. As a demonstration, this always attracts interested viewers.

A good species to use is a green frog or a leopard frog. Both kinds are common in ponds throughout the country. If proper conditions are provided, these frogs can be kept indefinitely in a schoolroom aquarium.

Place a medium-size frog in a battery jar or small aquarium and add some ice cubes to the water. If the demonstration is to be done as a science project and observers are present, a supply of ice cubes can be kept in a plastic foam container. Place a thermometer in the water and stir the ice cubes with it. As the temperature drops, the frog comes to the surface for air less and less frequently. At about 41°F. all motion usually ceases and the frog is in a hibernating condition.

Now remove any ice cubes that remain and allow the water to come back to room temperature. At what temperature does the frog rise to the surface? (If the water temperature is changed gradually the frog is not harmed in any way.)

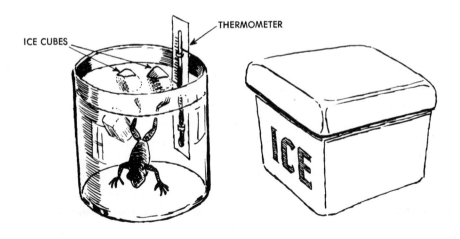

THERMOMETER

ICE CUBES

ICE

With a larger aquarium you can do an interesting variation of this experiment. Bend a sheet of metal and place it in the bottom of the aquarium as shown in the diagram. Fill the aquarium, place a frog in it, and lower the temperature as before. Under the sheet of metal it will be much darker than in the surrounding water. What will happen?

Observe the pulselike throat movements of the frog after it revives and comes to the surface. When a frog is in a hibernating condition under water it naturally cannot breathe air with its lungs. At such times, oxygen exchange takes place through the frog's skin. This also happens to some extent when a frog is in water above hibernating temperature. But when a frog is active, lung breathing is necessary in order to release sufficient energy for movement.

While you have a frog available, find the percentage of time that a frog is at the water surface compared with the time that it is under water at room temperature. Then raise the temperature by adding warm water and again find this percentage. Is it notably increased? What hypothesis can you suggest to explain this?

It might be interesting to repeat these experiments with two or more frogs. Do the frogs assume the same position when they are in a hibernating condition in water? These same experiments can be carried out with newts.

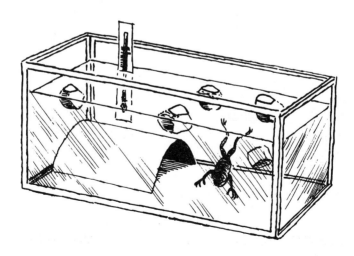

● Finding True Directions

It is possible, in a few thousand well-chosen words, to explain how to determine true north-south-east-west directions from the sun. But after reading or hearing these words, very few people would understand how it is done. Sometimes we need a working model to clarify even the simplest ideas.

Since the sun is to be used to determine true directions, it might seem impossible to make a working model to show how to find true directions from it. However, this problem is easily solved by means of an artificial sun. In the diagram here, the artificial sun is the bulb from a two-cell flashlight fastened to a hinge so that it sweeps across the "sky" as the real sun does.

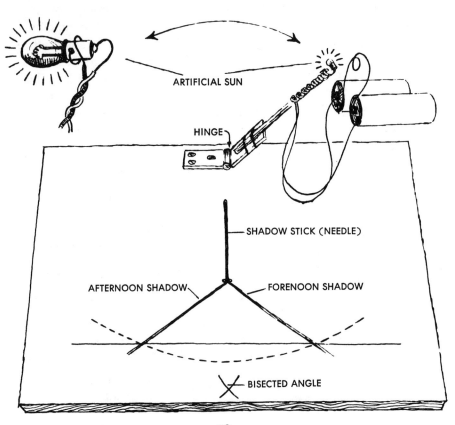

ARTIFICIAL SUN

HINGE

SHADOW STICK (NEEDLE)

AFTERNOON SHADOW   FORENOON SHADOW

BISECTED ANGLE

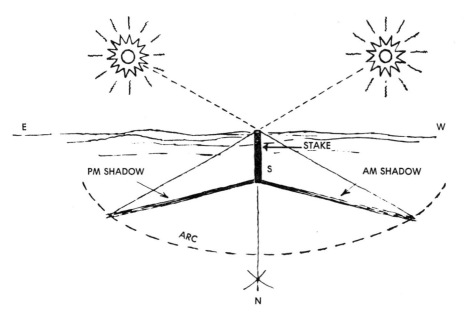

Select a point several inches in front of the hinge and use a compass to draw an arc (broken line) with this point as the center. Set up a long needle vertically at this point, turn on the sun, move it from left to right, and mark the points where the tip of the shadow crosses the broken line. As everyone knows, the sun is always directly in the south every day when it reaches the highest point in the sky.

When the tip of a forenoon shadow crosses the arc, this shadow will be exactly the same length as the afternoon shadow. Therefore, each shadow will be at the same angle with the north-south line. So all we need to do to find the true north-south line is to bisect the angle between the two shadows.

This can be said in several ways, but it will be much clearer if you set up the model and do the project yourself. According to the diagram, how are true east and west directions determined, and why?

Now you should be able to determine true directions anywhere on earth except near the equator. All you should need are two sticks and some string or rope. Explain.

• Demonstration Cold Front

Jack Pine had a good idea of a way to show the movement of a cold front. His plan was to take two glass window panes, separate them with small blocks, and enclose them with a metal strip so as to make a narrow box. Inside the box he would pour some oil and an equal amount of colored water.

To make the box watertight, Jack planned to use aquarium cement. But how was he to get the oil and water inside the box? Do you have the solution?

After Jack had solved this problem, he turned the box on one side. As expected, the water, which represented cold air, flowed under the oil, which represented warmer air. But this happened much too quickly. Something had to be done to slow down the movement. How might this be done?

When this problem had been solved, the water moved slowly across the model. But the thought occurred that the model did not show turbulence, as occurs when a cold front moves in. Perhaps some floating particles would be an improvement. What material might work best?

One day the model was left in a warm room and it sprang a leak due to expansion. How would you solve this problem?

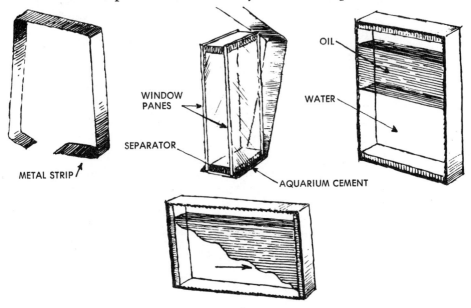

METAL STRIP

WINDOW PANES

SEPARATOR

AQUARIUM CEMENT

OIL

WATER

• Experiment with Germinating Seeds

After Jack had solved the problem described on the opposite page he started another project. He knew that radish seeds would germinate when placed on moist blotting paper. But would they sprout in the absence of oxygen? How would too much carbon dioxide affect germination? Would an excess of oxygen affect germination and growth?

Jack planned his experiment carefully. The materials he collected were four plastic refrigerator jars with lids, four large-size plastic pill bottles, twelve disks of blotting paper to fit the bottom of the refrigerator jars, several different chemicals, and of course some radish seeds.

First, Jack dipped the disks of blotting paper in water and placed three in each jar. Next, he filled the pill bottles as follows:

Pill bottle No. 1—This was nearly filled with dry sodium hydroxide, which absorbs carbon dioxide.

Pill bottle No. 2—A strong solution of pyrogallic acid with a little potassium hydroxide added was placed in this bottle. This mixture absorbs oxygen.

Pill bottle No. 3—This bottle was nearly filled with 20-volume hydrogen peroxide (the kind used to bleach hair), and a little manganese dioxide was added to release oxygen.

Pill bottle No. 4—Water only.

Now the bottles were labeled and set in the refrigerator jars. Radish seeds were scattered in the bottom of each jar and the lids put on.

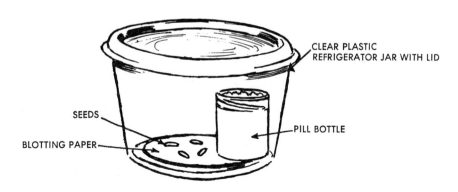

SEEDS

BLOTTING PAPER

CLEAR PLASTIC
REFRIGERATOR JAR WITH LID

PILL BOTTLE

• Topographic Models

A clear understanding of the meaning of the contour lines on topographic maps can be gained by building models. Make a miniature by moistening a gallon or two of ready-mixed concrete or by mixing a little cement and water with some sand. Pile up the moist mixture in the shape of a mountain.

After the mountain is dry and hard, get some half-inch or three-quarter inch pieces of waste lumber. Lay a pencil over one piece and move it around the base of the mountain, drawing a line on it as you go. Do the same thing with the pencil on two blocks, and continue adding blocks until you reach the top of the mountain. The result will be a three-dimensional model of a portion of a typical section of a topographic map. Every point on each contour line will be the same elevation above sea level.

If you should choose this topic for a science project, a topographic map of the local quadrangle will naturally be an essential part of it. These are available at a cost of about thirty cents per sheet from the Chief of Distribution, United States Geological Survey, Washington, D.C. 20025. A free copy of the Index Map for your state, showing the numbers of the quadrangles, can be obtained from the same source.

Study the topographic map and become familiar with the symbols used to indicate different features. Note the highest point, the lowest point, the contour interval, and other information.

Until a few years ago, contour mapping had to be done by surveyors and progress was slow. Now mapping is done by instruments carried by airplanes, and most of the United States has been mapped.

The diagrams show a second method of constructing three-dimensional topographic models. Place modeling clay in a large vessel and shape it in any way desired. Select a convenient contour interval, such as one-half inch equals x feet, and add a quarter inch of water measured by a ruler. With a dull pencil, mark around the clay at the water line. Continue adding water, a quarter inch at a time, and drawing in the contours until they are complete.

Assume that the base of the model is at sea level. Indicate the elevation of the various contours by markers of some kind, in accordance with the contour interval selected.

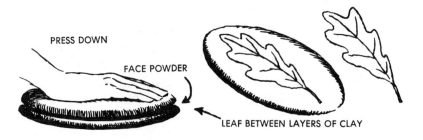

PRESS DOWN

FACE POWDER

LEAF BETWEEN LAYERS OF CLAY

● Artificial Fossils

Most natural fossils are formed when the remains of a plant or animal are covered with mud. Because most muds contain cementing materials, the rock particles stick together after a time and the imprint of a plant or animal is then sealed inside.

Modeling clay, which is really a kind of artifical mud, provides a quick way of showing how this can happen. Flatten a piece of modeling clay, dust some face powder on it, and lay a leaf over it. Dust it again and press more modeling clay down over the leaf. Then separate the two layers of clay.

Use plaster of Paris to make permanent artificial fossils. It is inexpensive and is sold in all hardware stores. Mix it with some water in a used ice-cream carton so that it is about the same consistency as melted ice cream. Add something to color the mixture to make it look like rock. Pour a layer in the lid of the carton, and when this hardens rub Vaseline over it. Lay the leaf on top, veiny side up. Rub it with Vaseline and add more plaster of Paris.

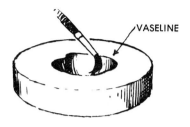
VASELINE
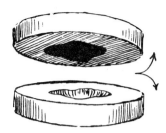

In museum exhibits the more unusual specimens are often copies of real fossils. One method of making such copies is shown in the diagram above. The real fossil is coated with a thin layer of Vaseline and pressed into wet plaster of Paris. When the plaster of Paris hardens, the surface around the specimen is coated with Vaseline and more wet plaster of Paris is poured on. When this hardens the two imprints are separated and the fossil is removed. Then both pieces containing the imprints are coated with Vaseline, and wet plaster of Paris, colored like the real fossil, is poured into each imprint and the surface of each is smoothed. When dry, the two halves are stuck together with an adhesive.

Pour wet plaster of Paris into animal tracks and make positive casts of them by the methods already described. Paper weights containing fossil imprints are interesting and easy to make.

The collection of genuine and artificial fossils is an excellent hobby and one that provides materials for a good science project exhibit.

FERN FROND   FISH BONES   RACCOON TRACKS

A. FERTILIZER   *B*. SUGAR   *C*. EMPTY

• A Double-Blind Experiment

Before a new medicine is offered for sale, it is supposed to be tested carefully to make sure that it is safe and effective. This is not always done scientifically, but, when it is, a procedure known as the double-blind method is commonly used.

How does the double-blind method work? Two sets of patients, either people or animals, are selected. Both sets are as nearly alike as possible, and all have the same disease or ailment. One set is given pills containing the new medicine. The other set is given placebos, usually sugar pills, which of course should have little or no effect. After a time, the results are compared.

In order to make sure that the person in charge of the experiment does not influence the results, several others, usually doctors, are selected to give out the pills. Half the patients are given the new medicine and the other half are given placebos. The doctors do not know whether they are giving the real medicine or only placebos, and the patients do not know either. That is why this is called the double-blind method.

With the aid of several classmates, you can set up an experiment to illustrate this double-blind method.

First, get some clean washed sand, some grass seed of a kind that has small seeds, and a supply of plastic or plasticized paper cups. Punch a hole in the bottom of each of the cups. Fill the cups with sand, sprinkle grass seeds in each cup, and barely cover the seeds with more sand. Set all the cups in a tray of very shallow water until the grass plants appear.

Now get some gelatine capsules from a drugstore and fill some with house-plant fertilizer and some with sugar. Mark the fertilizer capsules with one color of paint and the sugar capsules with another color. Call one of these A and the other B, and the empty capsules C.

Give each of your assistant experimenters three cups of germinated seeds, and one each of capsules A, B, and C. Tell them to punch a hole about an inch deep in the center of each cup and to push one of the capsules into it. Mark on the outside of each cup whether it contains capsule A, B, or C. Keep the cups in shallow water for ten days and note the results. Do they all agree?

Since the grass seeds were small, the little plants growing in sand soon suffer from malnutrition. The drawing here was copied from a photograph. What seems to cure malnutrition in grass plants?

● Effect of Light on Flowering Bulbs

It is customary to keep narcissus, hyacinth, and daffodil bulbs in the dark for several weeks after they are planted indoors in pots. But is this practice necessary? Does it produce healthier plants, better root systems, and more attractive flowers? What effect does it have on the length of time between planting and flowering?

To find out, obtain six similar bulbs and keep three in the dark and three in the light after planting. They may be planted in pots of soil or set among pebbles in a bowl. Maintain the same conditions except for the light. Keep the bulbs moist and well-

PLACE IN DARK   PLACE IN LIGHT

ventilated. When leaves appear on the plants that were kept in the dark, set them in the light next to the others and compare them from time to time thereafter. Which plants seem healthier? Which have the better root systems? Which ones bloom first? Which have the larger flowers?

With three more bulbs, you might set up a more elaborate experiment. Plant the three extra bulbs in the same way as the others but keep them in continuous artificial light twenty-four hours per day. For an exhibit, set them side by side with the story of each.

• MECHANICS AND TECHNOLOGY

The projects in this section should have a special appeal to those who already have some knowledge and like to prepare exhibits and models that make science ideas clear to others.

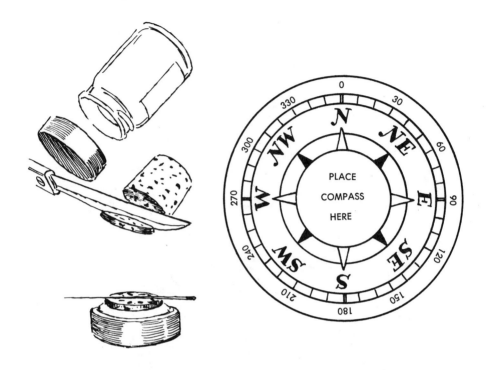

● Supersensitive Compass

Find a pill bottle and a cork that is slightly smaller than the cap of the pill bottle. Remove the cap from the pill bottle and slice off a thin piece of cork. Place the pill bottle on a flat surface and add water slowly until it is slightly overfull and the edges of the water are curved. Carefully drop the slice of cork on the water and place a magnetized needle on it.

Test the compass by bringing magnetized and unmagnetized objects near it. Compare its sensitivity with commercially made compasses. Why is this one almost frictionless? Look up "surface tension" to explain why the edge of the water in the lid is curved, and why the cork does not rub against the edge of the lid.

With the aid of a protractor, make a compass rose as shown here. Make it large enough so that the lid fits the space in the center.

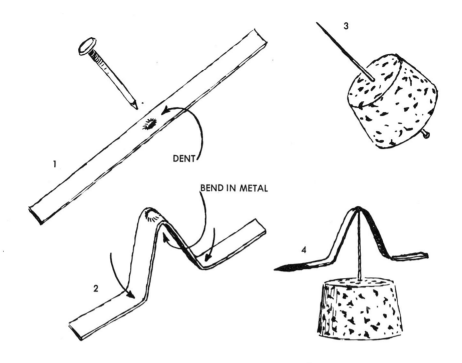

● Another Compass You Can Make

Here are some suggestions for making the compass shown in the step diagram:

Step 1—Cut the rims from an empty food can with a rotary can opener, and carefully cut a strip of metal from it with an old pair of shears. Stroke the strip twenty times in the same direction to magnetize it. With a blunt nail, make a dent (not a hole) at the center of the strip.

Step 2—Bend the strip at the dent and make two more bends as shown.

Step 3—Cut a slice from a cork and push a pin or needle through the center.

Step 4—Place the dent on the sharp point and snip off pieces from the ends until the strip balances. Mark the end that points north with a bit of paint. Test the compass by placing another compass near it.

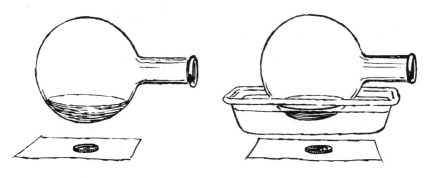

• Liquid Lenses

We see through liquid lenses. The watery sections in the front part of our eyes are shaped almost like the liquid in the left diagram above.

To show what happens when light from some object passes through such a convex liquid, put a little water in a boiling flask and hold it over some object such as a penny or dime. Flasks of this kind are usually included in chemistry sets.

To show what happens when light from a similar object passes through a concave water lens, lay the boiling flask in a shallow glass pan with a flat bottom and add some water to the pan.

The diagrams show how the light rays are bent (refracted) when they pass through each of the liquid lenses. Look up refraction of light in science textbooks and be able to explain the diagram.

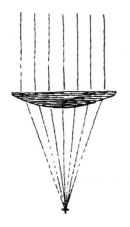
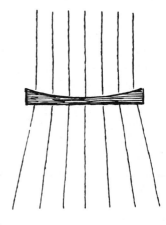

Wrap a thin wire around a medium-size nail and twist over the end to form a loop. Dip the loop in water and you will have a small double convex lens. Hold this loop over fine print and note what happens.

This principle can be used to make a microscope that will magnify up to 100 times: Cut five or six strips of aluminum foil and paste them together with rubber cement or other adhesive. Lay books on them to flatten them out as shown in A below. Now punch a hole through the center with a nail as in B and smooth the ragged edges.

Place the aluminum strip on a pane of glass supported above a mirror that reflects light upward. Dip a pencil in water and place a drop in the hole. Place something underneath that you want to examine, such as a hair, prepared microscopic slide, or a drop of water from a culture containing microorganisms (C).

If you are planning an exhibit, some glass lenses should be added for comparison.

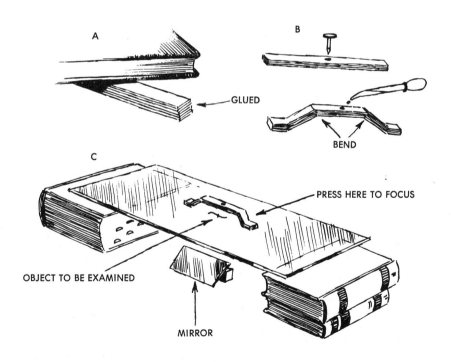

 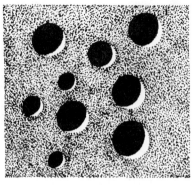

• Exhibit of Optical Illusions

In the picture at the left, which dish of ice cream looks larger? At the right, do you see holes or raised places? Turn the book upside down and look again.

In the two diagrams below, the grayish areas sometimes appear as the outer surfaces of the boxes. When you look again, they seem to be on the inner surfaces. Sometimes they change as you look at them.

Examples of this kind are called *optical illusions*. They show that our eyes cannot be trusted when scientific precision is needed.

Because of optical illusions our eyes fool us every day. A person wearing vertical stripes looks thinner than one of the same size wearing horizontal stripes. A white house looks larger than one of the same size painted a dark color. When we look down a road, it seems to become narrower in the distance.

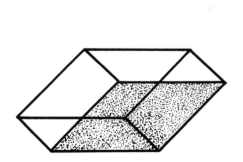 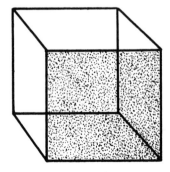

 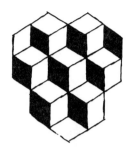

At the top of this page, the blocks at the left somehow look crooked and slightly different in size. At the right, you may see six blocks when you first look. If you look again you may see seven.

Below, the line AB seems shorter than the line BC. But if you measure them you will find they are exactly the same length. On the right, do you seem to be looking up from below the figure or looking down from it?

A collection of optical illusions would make a most interesting science exhibit. You can collect some using the library copier. Or you can redraw them quite easily if you make them in a larger size. Better still, invent some of your own. The examples here should give you some ideas.

It is normal for everyone to see optical illusions of this kind. But there is another kind that is commonly seen by people whose minds are disturbed by drugs. Find some of these "psychedelic" pictures and add them to your exhibit with appropriate captions.

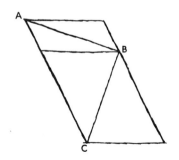 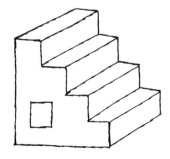

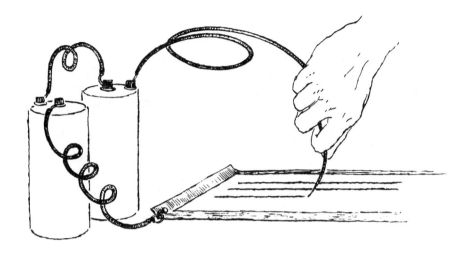

• Visualizing A-C and D-C

Many people do not understand the difference between alternating current (A-C) and direct current (D-C). Among them may be visitors who stop by to look at your science project.

It is always easier to explain something you can show than to depend upon words alone. So here is a way you can show how A-C differs from D-C.

Mix some cornstarch and a little potassium iodide in a small amount of water. Potassium iodide is included in most chemistry sets, and it also can be obtained from a school science laboratory or a drugstore. Soak a piece of white cloth in this mixture, squeeze it out, and smooth it over a cookie sheet or other shallow metal pan.

Now connect two wires to the terminals of one or two dry cells and attach the negative terminal to the pan or cookie sheet. Draw the end of the other wire across the wet cloth to produce a continuous dark line. Lift the wire and draw another dark line. Draw the end of a similar wire that is not connected with anything across the cloth. Nothing happens. This shows that electricity must be flowing continuously through the wire from the cell, and that it has something to do with causing a chemical reaction on the cloth.

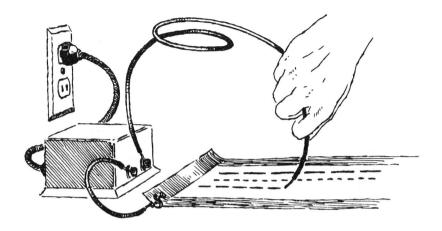

To illustrate A-C, use a transformer from an electric train set or a doorbell transformer. *Warning: Do not use 120-volt current directly from a main outlet because this would be very dangerous.* But with a small step-down transformer, this demonstration is no more dangerous than operating a toy electric train or ringing a doorbell.

Connect wires to the low-voltage side of the transformer and attach one wire to the pan or cookie tin. Now plug the cord into the outlet and draw the free wire across the cloth, as was done for D-C.

This time you should get a broken line or a series of dots, depending upon the speed at which the wire is drawn across the cloth. The dots and dashes appear only when the current from the wire is flowing in the positive half of its alternating cycle.

To further explain alternating current, you might make use of a diagram showing sine-waves, with explanatory labels.

Another addition to the demonstration might be to have a lamp handy wth a bulb that has the proper voltage to operate on the transformer. Also have a battery that delivers the same voltage as the transformer. Connect the lamp in one circuit and then in the other. Explain what happens in the filament of the lamp when it operates in each circuit.

● Electric Questioner

Testers for matching tests are usually quite complicated and require many hours or days of work. However, here is one that can be made quickly from a flashlight cell, a flashlight bulb, rubber bands, and the cardboard back of an 8½ × 11 writing tablet.

Lay a ruler about one inch from the edge of the cardboard and punch ten holes, one at every inch mark. Punch similar holes along the other edge. Remove the insulation from ten pieces of bell wire and connect the holes along the left side with those on the right, but crisscross the wires so that they are well mixed up. Bend the wires over the edges and fasten them by twisting as shown in *A*.

Turn the card over and straighten the wires as at *B*. Now fasten a long piece of bell wire to a flashlight bulb by making a spiral in the end as shown at *C*. Bend the wire so that the tip of the bulb touches the center terminal of the cell. Fasten another

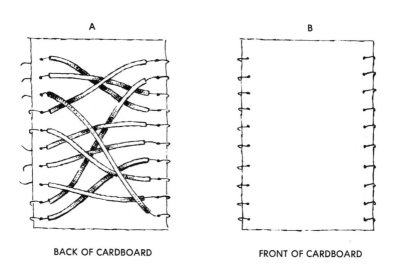

BACK OF CARDBOARD   FRONT OF CARDBOARD

wire to the bottom of the cell with a rubber band. Make loops at the ends of the long wires and test the circuit by touching the loops together. If the bulb lights, the tester is ready to operate.

To construct the matching test, make sure that the items on the left are properly connected with the answers on the right, depending upon the way that the wires are crisscrossed. Copy the test on a narrow piece of paper with the lines an inch apart so that they match up with the wires. Clip the test on the cardboard with paper clips.

The sample test shown here involves the ability to match electrical symbols with their meanings. The diagram shows that the bulb lights when "wires crossing" is connected through the cell and bulb with the symbol that has this meaning. The same is true for all the other items on the test. Any number of ten-item matching tests can be made up in a similar way and clipped to the cardboard. Tests in history, mathematics, and word synonyms are suggested.

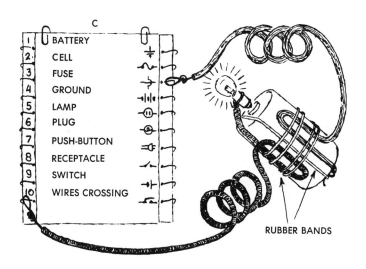

● Electrified Pendulum

In a clock that requires winding, the pendulum is kept swinging by the release of a little energy each time the pendulum makes a complete swing back and forth. This energy is stored up on the spring when the clock is wound.

Could electrical energy be used instead of mechanical energy to give the pendulum a little push so that it keeps swinging? The diagram on the opposite page shows one way this can be done. Many advertising signs we see in store windows operate on this principle.

The sketch at the bottom of this page shows an application of the electrified pendulum principle. This is interesting, but not particularly useful. Perhaps you can think of a better application.

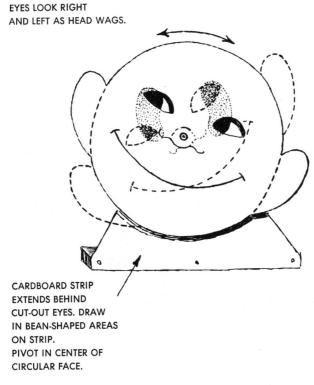

EYES LOOK RIGHT
AND LEFT AS HEAD WAGS.

CARDBOARD STRIP
EXTENDS BEHIND
CUT-OUT EYES. DRAW
IN BEAN-SHAPED AREAS
ON STRIP.
PIVOT IN CENTER OF
CIRCULAR FACE.

The diagram will help in building a gadget of this type.

- Points marked S on the diagram must be spot-welded or carefully soldered.
- If the permanent magnet is much longer than one inch, saw off a piece.
- The permanent magnet must be insulated from the wire that supports it. Try sealing wire for this.
- Two hundred and fifty turns of the enameled magnet wire is about right for the coil.
- To reduce friction, the "flexible metal strip" must be very flexible.
- Note that one lead from the coil is connected with the cell and the other is grounded to the metal support.

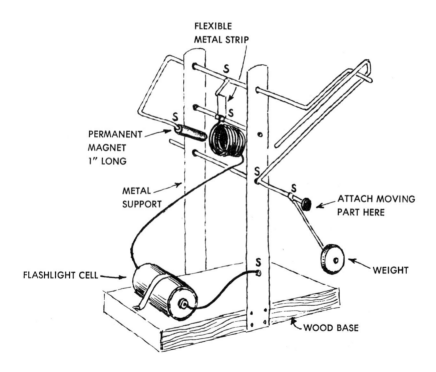

FLEXIBLE
METAL STRIP

S

S

S

PERMANENT
MAGNET
1" LONG

S

S

METAL
SUPPORT

ATTACH MOVING
PART HERE

S

FLASHLIGHT CELL

WEIGHT

WOOD BASE

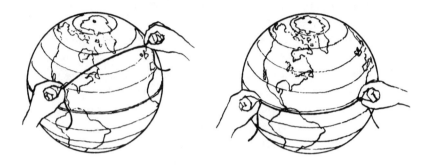

• A Milepost You Can Make

When traveling by car or bus, people like to watch for the mileposts that tell how far it is to the next large city or to the end of the journey. When traveling by plane it is interesting to know how far it is to the place you are going.

With a globe, a piece of string, and pencil and paper there is an easy way to find the distance between any two places on earth, assuming that you go by the shortest route.

First, take the piece of string and stretch it between the starting point and the destination. Hold both points of the string firmly and stretch the string along the equator. Begin at the point where the zero meridian crosses the equator in the Atlantic Ocean, just off the coast of Nigeria. Measure westward and note the number of degrees to the other point on the string. Multiply this figure by 69 and you should have the approximate number of miles between the two points first measured.

Why does this method work? The number of miles around the earth at the equator is about 25,000, and there are 360 degrees of longitude.

$$25,000 \div 360 = 69 \text{ miles per degree}$$

Suppose, for example, that you want to find the shortest airline distance between Philadelphia, Pennsylvania, and Tokyo, Japan. You measure this distance on the globe with a string, lay the string along the equator, and find that the total great-circle distance comes to about 140 degrees, some of it in west longitude and some in east longitude. Then you multiply 140 by 69 and get the answer, 9660 miles.

Now you are ready to start making your milepost, except for getting the directions, which you can do by looking at maps. Letter the signs and nail them to a firm support as shown here. Every school should have one. If entered in a science fair, the procedure used should be part of the exhibit. Place a magnetic compass beside it. Make sure that all signs point in the right directions.

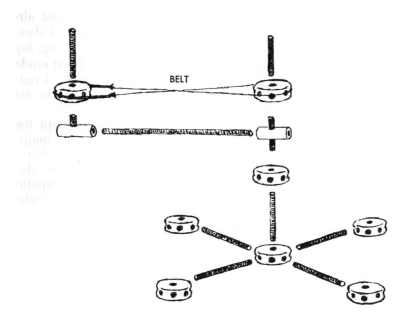

• A Planetarium You Can Make

A homemade or school-made piece of science apparatus is often more effective than one that is expensive and more complicated. For example, an excellent orrery, usually called a planetarium, can be made from a few Tinkertoy parts, an electric lamp, a needle, a piece of string, a rubber band, and two dime store plastic balls.

- Assemble Tinkertoy parts as shown in the diagram above.
- Fit an electric lamp and socket over the right-hand upright dowel.
- The ball representing the earth should be about the size of a baseball and made of lightweight plastic.
- The moon is a Ping-Pong ball, held in place with a long needle. Draw a face on the side facing the earth.
- The belt is a piece of string held taut with a rubber band stretched between the ends. Cross the belt so that the earth and moon both turn counterclockwise.
- Line up the "sun," "earth," and "moon" by sliding the earth upward on the dowel the right distance.

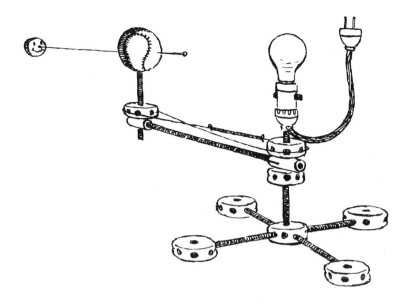

You may want to keep the planetarium for future use or exhibit it at the school science fair. If so, take it apart, dip the ends of the dowels in a good adhesive cement, and reassemble it. Among the demonstrations that can be carried on with the planetarium are:

- Counterclockwise revolution of the earth about the sun.
- Rotation of the earth in orbit around the sun.
- Counterclockwise revolution of the moon around the earth.
- Counterclockwise rotation of the moon as it travels in its orbit around the earth.
- The fact that the same side of the moon faces the earth at all times.
- The fact that the moon rotates on its axis each time that it revolves about the earth.
- Phases of the moon as seen from the earth.
- Length of day on the moon.
- Cause of day and night.
- Eclipses of the sun.
- Eclipses of the moon.
- The fact that the moon rotates on its axis.

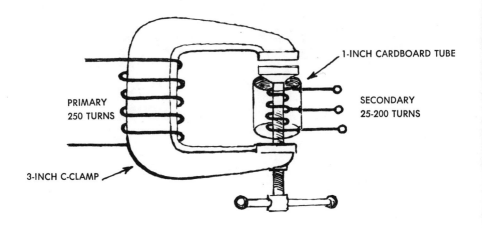

PRIMARY
250 TURNS

1-INCH CARDBOARD TUBE

SECONDARY
25-200 TURNS

3-INCH C-CLAMP

● Transformer from a C-Clamp

You can make a working model of a transformer from a *C*-clamp and some enameled copper wire as shown in the diagram. Connect the primary to a 6.3-volt 1-amp filament transformer, which is in turn connected with 115-volt house current. This low voltage prevents the coil from overheating and avoids the danger of shocks.

Wind the secondary around a cardboard tube with taps as shown. With this arrangement many experiments can be performed. Connect a 1.5-volt lamp across the secondary, using the different taps. Place nails inside the cardboard tube and note the effect on the brightness of the lamp. Open and close the clamp and note the effect. Connect a voltmeter in the circuit as shown in the diagram.

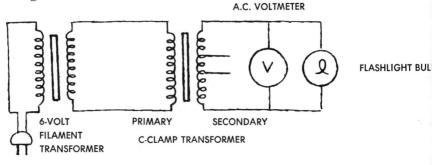

A.C. VOLTMETER

FLASHLIGHT BUL

6-VOLT
FILAMENT
TRANSFORMER

PRIMARY   SECONDARY

C-CLAMP TRANSFORMER

115 VAC

96

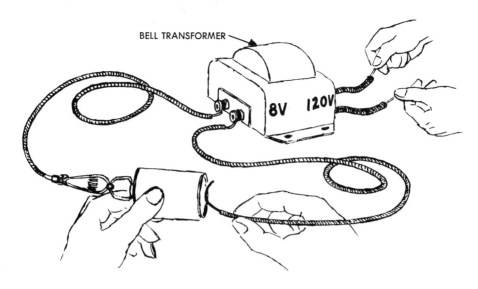

BELL TRANSFORMER

8V 120V

• Experiments with a Bell Transformer

In regular use, transformers operate only on alternating current. But it is interesting to connect a flashlight cell to the low-voltage terminals of a small transformer (the kind used to ring doorbells) as shown in the diagram above.

Ask someone to grasp the leads from the high-voltage side of the transformer while you touch the bottom of the cell with the free wire. A slight tingle should be felt when the circuit is closed and again when it is opened. Can you explain this? If not, ask someone who might know the answer.

This experiment should work better if the fingers holding the wires are moistened. Why? If a sharper tingle is wanted, use a No. 6 cell and connect a nail file as shown. Draw the end of the free wire across the file.

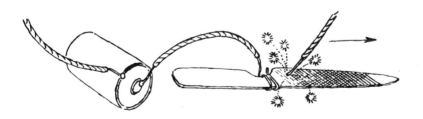

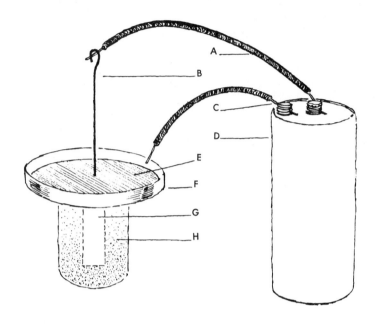

• Magic Motor

A. Lead from positive terminal (Bell wire).
B. Thin copper wire, loosely hung from A.
C. Lead from negative terminal (Bell wire).
D. Dry cell.
E. Thinnest possible layer of mercury.
F. Cardboard or plastic lid from container.
G. Alnico bar magnet; one pole touches bottom of F.
H. Jar filled with sand concealing G.

Adjust A so that B barely touches E. When C is depressed so that it barely touches E, B should describe circles in E. This happens because the magnetic field around B reacts with the magnetic field of G.

With the connections shown in the diagram, will the lower end of B whirl clockwise or counterclockwise? What will happen if A and C are connected to the opposite terminals of D?

• Build A Real Motor

The diagram below shows the main features of a toy motor that will run indefinitely if it is constructed carefully. Here are some suggestions:

- The rotor shaft is a pencil sharpened at both ends.
- The rotor shaft is supported by a strip of metal bent upward at each end and fastened to a wood base. A dent is made in each end for bearings.
- The armature consists of two nails taped together on opposite sides of the pencil and held in place by the two halves of a spool that has been sawed apart.
- The two armature segments are pieces of thin copper sheet bent around a second spool, which is slipped over the pencil armature shaft. These may be glued on or even held in place by rubber bands. Connect each of these to the ends of the armature winding.
- Cores of the field coils are nails. Windings around these end in the brushes that rest against the commutator. All windings should be fifty turns of enameled copper wire, and the directions of the windings are very important.

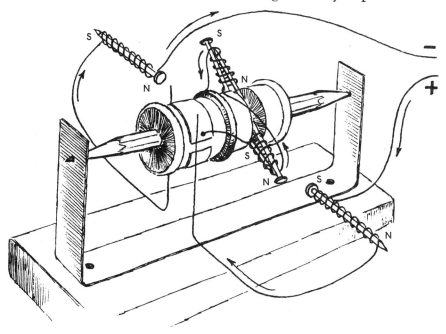

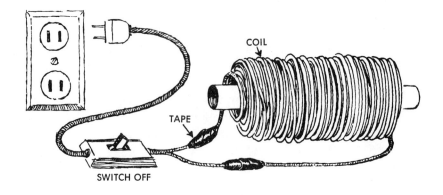

COIL

TAPE

SWITCH OFF

- **Instant Magnetizer**

Warning: This project is safe it it is carried out as described. After the apparatus has been assembled, the time required is only a fraction of a second. If a longer time is used, you may burn up the coil or blow out a fuse.

Find a cardboard tube about one inch in diameter and about ten inches in length. Or you can make one by rolling up a sheet of tagboard and tying or pasting it together. Wrap at least 500 turns of insulated wire around it and fasten the ends securely to a piece of extension cord, switch, and wall plug, as shown in the diagram. Cotton-insulated bell wire is satisfactory for winding the coil; lacquered wire should *not* be used. If splices are made, these should be carefully taped.

Now collect some iron rods or bars to be magnetized. If they are too long for convenience, saw them in pieces with a hacksaw. When everything is ready, insert one of the rods inside the coil. CHECK THE SWITCH TO MAKE SURE IT IS IN THE OFF POSITION. Then insert the plug in a wall socket.

Flip the switch on and off as quickly as you can, remove the plug from the wall socket, and remove the rod from the coil. It should be a strong permanent magnet.

As you know, electromagnets ordinarily operate on direct current. Ask someone who has a good knowledge of electricity why A-C works in this case.

● Magnetic Puzzle

Obtain two rods that are alike except that one is a permanent magnet and the other is not. The magnetized rod can be made as described on the preceding page.

Hand both rods to someone and ask the person to tell which one is a magnet and which is not. Few people seem able to solve this puzzle, so we will let you in on the secret. Let us call the magnetized rod A and the unmagnetized rod B. The following statements refer to the diagrams below:

(1)   Both ends of rod A attract rod B.
(2)   Both ends of rod B attract rod A.
(3)   Rod B does not attract the center of rod A because this center is midway between the poles, and therefore its magnetic force is neutral.
(4)   Rod A attracts the center of rod B and therefore A must be the magnetized one.

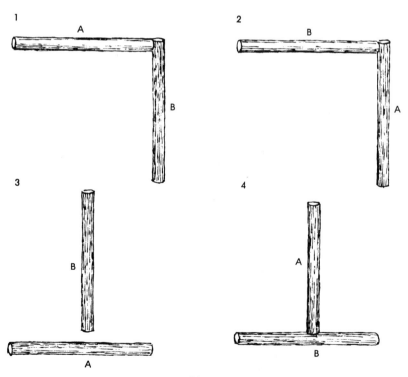

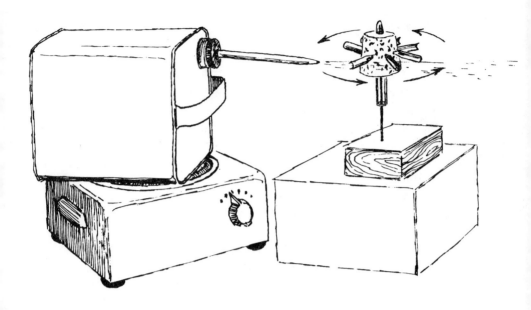

• Working Model of a Turbine

The rotor for a model steam turbine can be made from a short piece of glass tubing, a cork, a few penpoints, a long sewing needle, and a block of soft wood for a base. Heat the piece of glass tubing in a hot flame until the end melts and closes. Now make a hole through the center of the cork and press the closed end of the tubing through it. Stick penpoints in a circle around the cork to serve as blades for the rotor. Push the eye end of a long sewing needle into a block of soft wood and set the rotor on the point of the needle.

For a boiler, get a gallon metal can and a one-hole rubber stopper to fit the opening. Into this push a piece of glass tubing, the protruding end of which has been drawn out to form a smaller opening.

Now partly fill the can with water, set it on a source of heat such as an electric hot plate, and set the turbine in the position shown in the diagram. Regulate the source of heat so that the jet of steam spins the turbine at a good rate.

Working Model of a Geyser

Canned heat is another convenient source of energy for working models requiring heat energy. In the diagram below, the purpose of the marbles in the coffee holder is to hold the inverted funnel firmly against the bottom of the pot. The purpose of the aluminum foil pie tin is to catch the hot water that comes from the geyser and return it to the pot.

Assume that the marbles represent the rocks on the surface of the earth and that the heat comes from the interior of the earth. If a glass perculator coffee pot is used for the model, a fairly good representation of the operation of a geyser can be shown.

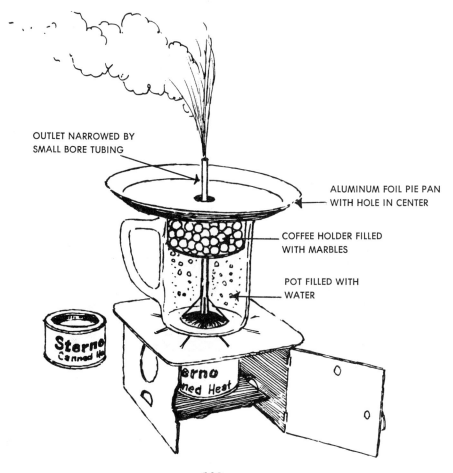

OUTLET NARROWED BY
SMALL BORE TUBING

ALUMINUM FOIL PIE PAN
WITH HOLE IN CENTER

COFFEE HOLDER FILLED
WITH MARBLES

POT FILLED WITH
WATER

• Fingerprints with Sticky Tape

In order to get clear fingerprints that are good enough for identification purposes, you should have a regular fingerprint outfit. Fingerprints can also be made with the aid of a stamp pad, but these are too smudgy to be of much use. However, there is a simple way to make excellent fingerprints with only a sheet of paper, a pencil, and a roll of clear cellophane sticky tape.

Rub a soft pencil at a slant back and forth over a sheet of paper to make a black area. Rub your finger over this and press it against a wide piece of clear transparent sticky tape. Press the tape on a sheet of paper and you should have a very clear print. The trouble is that the print is in reverse.

To make a fingerprint that the FBI could use, paste another piece of tape over the sticky side of the tape as you remove it from your finger. This piece must be longer than the other so that the ends will hold against the paper.

With a series of prints made in this way, make an exhibit showing the principles of fingerprint classification (arches, loops, whorls, and composition).

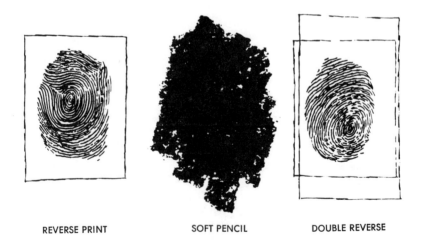

REVERSE PRINT   SOFT PENCIL   DOUBLE REVERSE

• Two Methods of Breathing

The apparatus shown at the left below has long been used to demonstrate the process of breathing. The rubber balloons represent the lungs, the Y-tube represents the windpipe and trachea, the bell jar represents the thoracic girdle, and the piece of sheet rubber represents the diaphragm. Lowering the diaphragm reduces the pressure inside the chest cavity and air flows into the lungs. Raising the diaphragm reverses the flow of air.

But the ribs are also important in breathing, especially during exercise. A simple apparatus can be set up to demonstrate this in terms of air pressure.

Clear or almost clear plastic bottles are available in many stores. Buy a large one that has a neck that fits a large one-hole rubber stopper. Cut off the branches of a Y-tube enough so that they can be inserted in the plastic bottle, and attach the balloons and stopper. The result will be similar to the apparatus on the left, except that there is no diaphragm and the sides are flexible.

Squeeze the sides of the plastic bottle to increase the pressure inside, then release. Exhibit the two devices side by side and explain them in terms of air pressure.

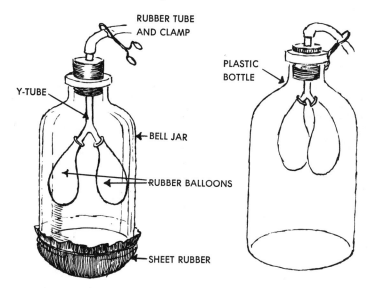

105

• Mock-Up of a Doorbell Circuit

How many of your friends understand what happens when someone presses the button of your doorbell? Do you?

For an interesting and instructive exhibit, get a large sheet of plywood, finished or painted on one side, and mount the circuit as shown in the diagram.

Press the front-door push-button and trace the path of the electricity through the circuit. Then press the back-door push-button.

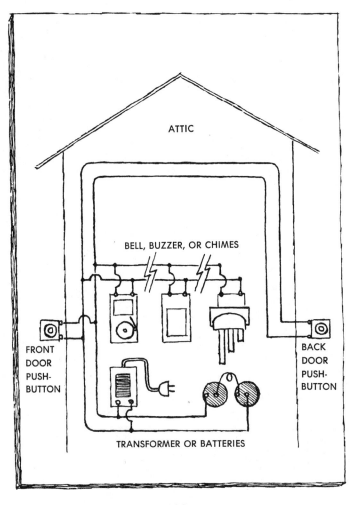

Mock-Up of a Three-Way Lamp Circuit

How many of your friends understand how it is possible to turn a lamp on at the bottom of a stairs and turn it off when you reach the top? Do you?

Make a mock-up of this circuit on a sheet of plywood, using two double-throw switches, dry cells, and a flashlight bulb in a small porcelain socket.

Will the lamp light with the switches in the positions shown? Now throw either switch, then the other.

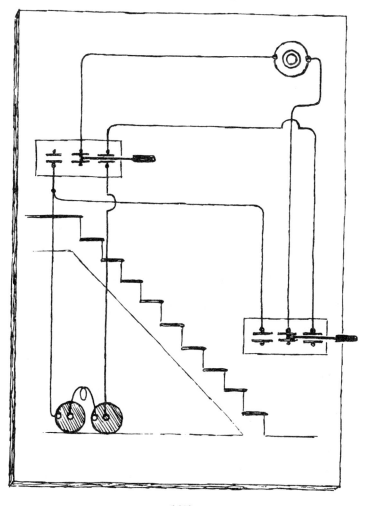

● Test Your Can Opener

Do you eat bits of metal shavings? To find out, save up six or more metal cans in which food is packed. Wash the cans as clean as possible and dry them thoroughly.

Now place a sheet of white paper on a support under the can opener and open the bottom ends of the cans. Examine the white paper under a good magnifying glass and look for tiny metal chips and shavings. Run a permanent magnet back and forth across the paper and examine the end of this with a magnifying glass.

Not many can openers can pass this test. It is not definitely known how harmful these bits of metal are when eaten, but they are probably not very nutritious.

For an exhibit, it would be instructive to set up a can opener and carry out this demonstration—with a plentiful supply of empty cans.

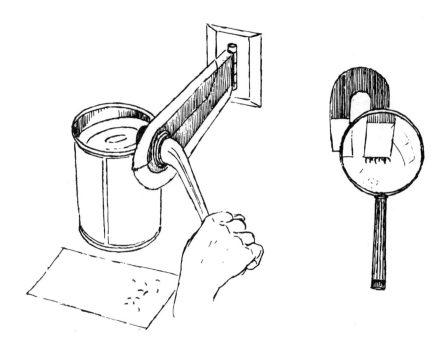

Cooling Effect of Moving Air

Three bottles are filled with water and a thermometer is placed in each. When the temperature of all three agrees with the room temperature, a wet towel is wrapped around bottle 2. Then a stream of air from an electric fan is directed toward bottles 1 and 2, which are both the same distance from the fan. Bottle 3 is placed where the stream of air does not strike it.

Record the temperature of the water in each of the three bottles in fifteen minutes. Wet the towel again and record the temperatures after another fifteen minutes. If four thermometers are not available, use one laboratory thermometer and hold it in each bottle for a few seconds in turn.

What conclusions can be drawn from this experiment?

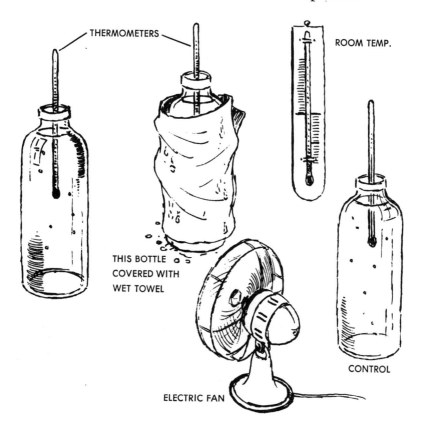

THERMOMETERS

ROOM TEMP.

THIS BOTTLE
COVERED WITH
WET TOWEL

CONTROL

ELECTRIC FAN

● Effect of Light on Molds and Mildews

In the "household hints" column of a leading women's magazine, the claim is made that if continuous light is provided, molds and mildews will not grow in a closed closet. This claim seems to be worth checking since much damage is done to clothing by molds and mildews.

It is easy to demonstrate that moisture is necessary for the growth of molds. The method outlined in many science textbooks goes something like this: Keep a slice of bread in the air until it is very dry, then moisten it with water, place it in a jar, and

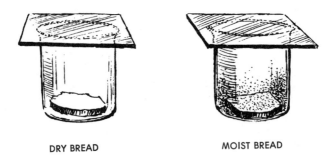

DRY BREAD   MOIST BREAD

cover the jar tightly. For the control, do exactly the same thing without moistening the bread.

For mildew, use pieces of wool cloth instead of bread. Put a bit of mildewed cloth in each jar, to insure that mildew spores are present in both. Moisten one but not the other.

In these experiments, the only thing that is controlled is moisture. What will happen if the jars are kept in bright light continuously for twenty-four hours each day instead of for about half that long? It is possible that no one has tried this form of the experiment to find out. Moisture might not be the only factor necessary for the growth of molds and mildews.

• INDEX

light, refraction of, 82
light beam, 19, 26
lines of force, 36–37, 41
longitude, 93
longitude, celestial, 57
loop (fingerprint), 104

magnesium sulfate, 11, 20–21
magnet (permanent), 36–37, 41, 91,
  98, 100–101
magnetic field, 36–37, 41, 98
magnetic field of earth, 40
magnetic force, deflection of, 37
magnetization, 40–41, 80, 100
magnification (magnifying power), 39
magnifying effect, 43
magnifying glass, 20, 65, 108
malnutrition, 77
manganese dioxide, 71
meridian (prime), 57, 92
meridian, celestial (prime), 56–57
microorganism, 38, 83
microscope, 38, 52, 83
migration, 49
milepost, 92–93
mirror, 18–19, 22–23, 26–27
moire pattern, 28
mold, 49
molecule, 20–21, 34
molecule, motion of, 30
moon, 94–95
moon, axis of, 95
moon, phases of, 95
moon, revolution of, 95
moon, rotation of, 95
motor, 98–99

naphthalene, 38
Newton's Second Law of Motion, 44
North Star, 56–57

optical illusion, 84–85
oxygen, absorption of, 71
oxygen, effect on germination, 71
oxygen, effect on growth, 71
oxygen, release of, 71
oxygen exchange, 67

parapsychology, 24
pendulum, 15, 90
petal, 64–65
pistil, 64
placebo, 76–77
planarian, 50
planetarium (orrery), 56, 94–95
polarity, 50
polarization (of light), 38
polarization (of worms), 50
polarized plastic, 38
potassium hydroxide, 71
potassium iodide, 86
propagation (of vegetation), 62
pyrogallic acid, 71

ray (of flower), 64
reflection of light See light,
  reflection of,
refraction of light See
  light, refraction of,
rhyzome, 62
rocket, principle of, 44
runner, 62

salt, 11, 17
salt, table See sodium chloride
seasonal change, 49
seasonal change, effect on
  animals, 49
seasonal change, effect on stars, 49
seasonal change, effect on spider
  webs, 60
sepal, 64
sine-waves, 87
sodium carbonate (washing soda), 17
sodium chloride, 17, 34, 42–43
sodium hydroxide, 71
solubility, 42–43
sound vibrations, 27
spider webs, 60
stalactite, 11
stalagmite, 11
stamen, 64
star (group), 54–57
statistical probability, 24
stroboscopic effects, 14–15
sulfuric acid, 17
sun, 55, 68, 94–95
surface tension, 35, 80

Taurus, 55
temperature, 33, 47, 66–67, 109
terminal (of dry cell), 50, 86,
  88, 97–98
thoracic girdle, 105
topography, 72–73
trachea, 105
transformer, 87, 96–97
transmission (of image), 25
turbine, 102
turbulence, 70

vascular bundle, 63
Vega, 57
velocity, 30
violet methyl, 42
voltage, 87, 96–97
voltmeter, 96
vortex rings, 12–13

water table, 58–59
weather conditions, effect of, 52, 60
wetting agent (detergent), 35
whorl (fingerprint), 104

Zener cards (ESP testing cards),
  24–25
Zodiac, 54–55